Quality in Science

Quality in Science

edited by
Marcel Chotkowski
La Follette

The MIT Press
Cambridge, Massachusetts
London, England

MIT Press

0262620405

LA FOLLETTE
QUALITY SCIEN

© 1982 by the Massachusetts Institute of Technology and the President and Fellows of Harvard College

All rights reserved. No part of this book may be reproduced in any form or by any means, electronic or mechanical, including photocopying, recording, or by any information storage and retrieval system, without permission in writing from the publisher.

This book was typeset by Grafacon, Incorporated, and printed and bound by Halliday Lithograph in the United States of America.

Most of the essays in this book first appeared in the Winter and Spring 1982 issues of the quarterly review *Science, Technology, & Human Values*. Publication in the journal was supported by the National Science Foundation (NSF) and the National Endowment for the Humanities (NEH) under grants OSS-7918701 and SRS-8007378 to Harvard University (G. Holton, Principal Investigator). Any opinions, findings, conclusions, or recommendations expressed are those of the authors or editor, and do not necessarily reflect the views of the NSF or NEH.

Special thanks are due to the staff of *Science, Technology, & Human Values*, especially Lisa Buchholz and Melinda Thomas, for their work on this book, and to the journal's Advisory Board for their support of the project.

Library of Congress Cataloging in Publication Data

Main entry under title:

Quality in science

"Most of the essays in this book first appeared in the winter and spring 1982 issues of the quarterly review, Science, technology & human values"—T.p. verso.

Includes bibliographies and index.

Contents: Foreword/by Gerald Holton—Science indicators and science priorities/Harvey Brooks — Needs, leads, and indicators/Robert S. Morison — [etc.]

1. Science indicators—Addresses, essays, lectures.
2. Quality (Philosophy)—Addresses, essays, lectures.
I. La Follette, Marcel Chotkowski. II. Science, technology & human values.
Q172.5.S34Q34   1982   501'.3   82-42598
ISBN 0-262-12099-2 (cloth)
ISBN 0-262-62040-5 (paperback)

Contents

Foreword vii	Gerald Holton
Science Indicators and Science Priorities 1	Harvey Brooks
Needs, Leads, and Indicators 33	Robert S. Morison
The Quality of 'The Quality of Science': An Evaluation 48	Bruce Mazlish
Industry Evaluation of Research Quality: Excerpts from a Seminar 68	Lewis Branscomb
The Public and Science Policy 82	Kenneth Prewitt
Changing Public Attitudes to Science and the Quality of Life: Excerpts from a Seminar 100	Daniel Yankelovich
The Social Assessment of Science, or the De-Institutionalization of the Scientific Profession 113	Peter Weingart

The Quality of Science Equation 119	Orrin G. Hatch
The Role of the Federal Government in Supporting Research and Development 122	George E. Brown, Jr.
Decision-Making for Quality Science 126	Don Fuqua and Doug Walgren
Secrecy and Openness in Science: Ethical Considerations 131	Sissela Bok
Ethical Issues in the Assessment of Science: Excerpts from a Seminar Series 149	

Afterword 157 — Marcel Chotkowski La Follette

Contributors 163

Index 165

Foreword   Gerald Holton

The authors of the essays in this book accepted the task to explore what old or new operational meanings might be associated with the concept of "quality," when applied either to the state of science and technology or to the impacts these have on human life. They ask whether—and, if so, how—indicators may be developed which are sensitive to the various contexts of science (conceptual, ethical, social, historical). What are possible measures of quality? What are the constraints on quality? How do different interested constituencies—scientists and engineers, the public, bureaucrats and foundation administrators, industry, the Congress—assess the quality of science, and what measures would be most useful to these groups?

These are questions to which, at this stage of the discussion, we cannot expect pat answers; what we need are proposals that will stimulate more detailed formulations and demonstrate the variety of approaches available for understanding the quantity-quality dichotomy. This effort has a place in a tradition going back to Aristotle and Nicole Oresme. Far from being an exercise for antiquarians, however, the attempt to determine indicators of quality responds to contemporary concerns. One index of these concerns is the list of distinguished contributors to this volume and the wide array of participants at the faculty seminar series during which the ideas for most of these essays were first presented for discussion.* The intrinsic interest of the subject assured the presence of scientists as well as scholars and practitioners involved in science policy, sociology of science, history of science, administration, and humanistic studies. The institutions represented included government, industry, and private foundations, as well as universities.

* It is a pleasure to thank the Program in Science, Technology, and Society, MIT, for the hospitality extended to the faculty seminar series held under its auspices.

On this topic converge many of today's efforts in the field of science resource studies—for example, current pressures for setting research priority policies; re-examination of traditional institutions and curricula for the training of scientists; the debate among sociologists and philosophers of science on the meaning of "scientific progress"; public discussion of the roles science and technology play in the lives of citizens; and the growing awareness among some scientists that, more and more often, "extra-scientific" social standards and ethical rules are used to evaluate the funding and conduct of scientific and technological activity, the choice and design of research problems, and the applications of research results.

Once alerted, one finds it difficult to escape noticing the persistence of questions about quality. Selecting almost at random from recent science publications on my desk, I note how often such phrases as "the health of science" recur: "Currently the NSF responsibilities for the health of science are being eroded by the Congressional appropriations process, initiated and spurred by the Office of Management and Budget" (Emanuel R. Piore, 1981); "Science and Government may be approaching a moment of decision in which the health of both is at risk" (Frank Press, 1981); "There [has] been an extraordinary decline in the quality of research equipment now available in the major universities" (Donald Kennedy, 1981); "No indicators are available to measure or predict the future quality of American scientists and engineers. ... There are indications that the U.S. higher education system is under considerable strain and is not able to provide a high quality education in science. ..." (NSF report on science and engineering education, 1981).

The word "quality"—or terms with the same meaning—also appears frequently in remarks by officials of the current U.S. Administration. For example, in December 1981, George A. Keyworth, President Reagan's science advisor, told the House of Representatives' Committee on Science and Technology, "My own experience leads me to believe that the best overall quality of research may not occur in time of accelerating support, but in times of moderate restraint." Six weeks earlier, he had explained to a group of Department of Energy physicists that the recent cuts in Federal funding of science and technology "provided a good opportunity to speed up the process of distinguishing between first- and second-rate research" (*Science and Government Report,* 15 October 1981).

Similar expressions of a desire for (or the assumption of) quality indicators appear throughout the literature—for example, the NSF study on the quality of the peer review system; the National Science Board's June 1981 statement on the social and behavioral sciences, which contained a call to enhance "the objectivity of the sciences and improve the quality of data collection and analysis" as well as "quality and usage of national statistical information"; the National Research Council's report on "Assessment of Quality-Related Characteristics of Research-Doctorate Programs in the United States" (1981); the concern of the Chemical Abstracts Service about the "quality of the original chemical information being processed and disseminated" (*Chemical & Engineering News*, 1 June 1981); and opinion research findings that between 1957 and 1979 there was a substantial drop in the percent of polled citizens who, in assessing the impact of science on their lives, believe that "on balance the benefits of scientific research have outweighed the harmful results" (*Science*, 15 January 1982). The universe of discourse reaches from these current citations to scholarly works such as that of Comroe and Dripps on the identity of high-quality biomedical science (1976). Even in the recent Arkansas "Scientific Creationism" trial, criteria were proposed for the qualitative earmarks by which a true work of science can be recognized ("explanation, prediction, testing, consistency, consilience").

Given the ferment, it is puzzling that quality indicators have so far not been the subject of more widespread and coherent study. In principle, they are part of the relatively well-launched study of "social indicators"; yet it has become increasingly evident that the particular social activity called scientific research, and other aspects of science and technology, have not been of major interest to those involved in the production of social indicators. In the report *Social Indicators 1976*, for example, data about science are so scant that the word "science" does not appear as a main entry in the index—unlike "television," "firearms," "homicide," etc. (Science is mentioned occasionally in sub-entries—e.g., "science achievement scores," which are listed strangely under the main heading "Residence.")

Such national assessments of science as have been attempted are mainly quantitative, and have not been designed to permit the deduction of clear qualitative conclusions from the quantitative data. Serious development of quantitative science indicators began only about a decade ago, with the publication of the National Science Board's *Science Indicators 1972*. In the meantime, substantial primary and secondary literature has accumulated through subsequent issues of *Science Indicators,* the volume *Toward a Metric of Science: The Advent of Science Indicators* (1978), the journal *Scientometrics,* and reports from meetings such as those sponsored by the Social Science Research Council and the Organization for Economic Co-operation and Development, to name just a few.

In all these efforts, the absence of qualitative efforts is virtually complete, although sometimes explicitly regretted. On the first page of the first issue of the National Science Board's *Science Indicators* series (1972), it was stipulated that intrinsic measures should include those of both "quantity and quality of associated human resources," and that extrinsic indices would center on "the achievement of national goals . . . and the consequent impact on that elusive entity, the 'quality of life'." Perhaps reflecting back on that ambition, a member of the National Science Board, during a May 1976 Congressional hearing on measuring and evaluating the results of federally-supported research and development, acknowledged that "the present effort to assess U.S. science [was] still only in the early stages of maturity," and that what is required is to "seek new measures of the state of science. It is hoped that all those interested in science indicators will participate in this search."

In fact, a small number of attempts have been made to develop such qualitative indicators — such as the three that appeared in the first issue of the *Science Indicators* series (the relative number of Nobel Prizes received by U.S. scientists, which is understood as a quality indicator; citation analysis, referring to the relative standing of a nation's literature with respect to the total literature in science; and "degrees of innovation" as rated by panels of experts). On reflection, a number of other, perhaps more plausible, candidates come to mind. One could experiment with developing measures of the quality of a field or a journal by specific peer review of a sample of the output, analogous to the Bell Laboratory assessment of journal articles in the field of condensed matter physics. Or one could examine philosopher John Passmore's proposal that the two chief measures of the quality of a scientific work are fecundity and synthesis. To avoid the danger of concentrating on aggregates, one could try to identify specific crucial events, including institutional as well as conceptual discontinuities or major advances on which relatively easy consensus might be reached. (This last proposal may be the chief reason for the inclusion in the most recent volume of *Science Indicators* (1980) of a

Foreword   xi

new chapter on "Advances in Science," which is introduced with the frank comment, "Most quantitative indicators fail to capture qualitative aspects of the entities they measure or do not describe the complexities and nuances of the processes involved. These limitations had become increasingly evident in the earlier editions of *Science Indicators* which covered quantitative input and output measures. . . .") In the following pages, the reader will find a great variety of proposals for carrying forward this widely desired but newly begun development.

Jefferson Physical Laboratory
Harvard University
2 April 1982

References

Julius H. Comroe, Jr., and Robert D. Dripps, "Scientific Basis for the Support of Biomedical Science," 192 *Science* (9 April 1976): 105–111.

Donald Kennedy, speech delivered 23 November 1981, 14 *Campus Report* 10 (Stanford University).

George Keyworth, quoted in *Science and Government Report* (15 October 1981): 8.

National Science Board, *Science Indicators 1972, Science Indicators 1974, Science Indicators 1976, Science Indicators 1978,* and *Science Indicators 1980*. (Washington, DC: National Science Foundation).

National Science Foundation, *Science and Engineering Education in the 1980s and Beyond* (October 1981): 60–64.

Emanuel R. Piore, "Physics Funding," *Physics Today* (December 1981): 45.

Frank Press, "New Cuts in Agency Budgets," 214 *Science* (16 October 1981): 261.

Michael Ruse, "Creation Science Is Not Science," 7 *Science, Technology, & Human Values* 40 (Summer 1982).

Quality in Science

Science Indicators and Science Priorities

Harvey Brooks

Introduction

This series of essays has been partially motivated by dissatisfaction with the science indicators that have so far been devised to assess the health of science in the United States. In many ways, this sense of dissatisfaction parallels the increasing disenchantment with economic indicators, and for the same reasons—that they represent a lumping of many different kinds of inputs or outputs. In the case of the economic indicator GNP, for example, expenditures to cure disamenities, such as those on pollution control equipment, are regarded as adding to the output already measured in the output of goods that generated the pollution in the first place. On the other hand, improvements in the quality of goods, such as improvements in durability, reliability, or energy consumption, may actually reduce output through reducing replacement production or expenditures on repairs or primary energy sources.[1]

In the case of science indicators, the difficulties are even greater because most indicators are input measures rather than output measures, a situation which obtains in the case of economic indicators for only a few sectors, such as government services. Where output measures have been introduced, they tend to measure the internal mechanics of the research enterprise itself—scientific papers, number of citations to scientific papers, patents, Nobel or other prizes, invited papers at international meetings, and sometimes migrations of scientists. Measures of the quality or significance of these outputs are crude at best. Moreover, neither the input nor the output measures give much weight to the social functions of the science enterprise, or provide any criteria for the assessment of its output against the social functions expected of it.[2]

Of course, any system of numerical indicators is bound to be crude and oversimplified. Economic indicators such as GNP have tended to acquire an illusion of objective measurement that few would judge as deserving once the process by which the indicators are actually calculated is revealed. Yet economic indicators have evolved over more than half a century, whereas science indicators are less than a decade old. All indicators involve an inherent tension between simplicity and reality, and some would argue that science indicators are no more oversimplified than many other kinds of social indicators, from labor productivity to health statistics. Most social indicators are designed to abstract, however imperfectly, some characteristic of society as a whole, whereas science indicators tend to focus on a single set of activities in a way which makes the set appear entirely self-contained, with no interface with the rest of society. GNP per capita, infant mortality, or average number of years of schooling purport to measure an attribute of a whole society, while R&D expenditures per capita, R&D as a fraction of GNP, or numbers of scientists and engineers per 10,000 workers are not so easily interpreted.

In this essay I shall be concerned primarily with science in its interface with the rest of society, and will try to indicate some of the difficulties and complexities involved in developing realistic measures of the state of science which can translate more self-evidently into indices of welfare for society as a whole. The approach will of necessity be largely conceptual. We appear to be a long way from being able to devise quantitative indicators to characterize this interface, or even qualitative indicators that can characterize more than particular pieces of the system.

Intrinsic vs. Extrinsic Indicators

Throughout the history of science as a social activity, a continuous tension—indeed a struggle—has existed between the autonomous goals of science, and its sponsors' or patrons' desire for results of demonstrable social utility.[3] However, this tension exists also between different parts of the technical community as well as between the technical community and its various government and industrial patrons. Most of the technical community is engaged in work which is managed directly in terms of social or economic objectives. Probably ninety percent of the technical enterprise is managed in this way;

indeed, if we take the use of peer review as a criterion for self-management within the scientific enterprise, then the fraction of research that is self-managed is probably less than five percent. This fairly obvious point bears emphasizing because theoretical discussions of R&D management too often appear to presume that the entire $60 billion investment, divided equally between private and public sources, is managed in the fashion of the investigator-originated research projects of the National Science Foundation or the National Institutes of Health, with the technical community operating under a "social contract"[4] that Vannevar Bush negotiated with the political system in 1945 and laid out in the report, *Science, the Endless Frontier*.[5]

To be sure, tension between intrinsic and extrinsic decision criteria in the management and organization of research exists even in the parts of the enterprise that are managed from the top down in terms of social and economic objectives. This is true because the ability to achieve nonscientific objectives with the aid of science is constrained both by nature itself and by the current state of our understanding of nature. Social results cannot be forced beyond a certain point if nature and science do not permit it, and it is the technical people who live closest to these constraints. In the words of Charles Fried, we cannot affirm "truth as a constraint on the pursuit of the useful, while denying truth any power to set its own agenda."[6] Thus, even the most applied researchers may come into conflict with their employers if nature does not appear to be cooperating with the goals, timetables, or policy presumptions on which the original research effort was predicated.

The same tensions that arise in the management of the technical enterprise also arise in its measurement. In assessing the health of the enterprise, how do we design indicators that reflect a suitable balance between elements internal and external to it? Extrinsic indicators would entail not only measuring progress toward certain social objectives, but also judging the relative worth of the objectives themselves. For example, we might design indicators of progress in military technology, since national security is certainly one of the social objectives that the political system has set for the technical enterprise. But many people are concerned that progress in military technology is one of the prime engines of the arms race, which may in a true sense represent a net subtraction from national security, not to mention the enormous opportunity cost to world development implicit in the devotion of such a large proportion of the world's technical talent to military R&D.[7] Should the indicator be a measure of national or international security in some fundamental sense, or merely some measure of unilateral military capacity? Indicators of military capability may be intermediate between purely intrinsic measures and truly extrinsic ones; they are easier to design, but they are less meaningful from the standpoint of a genuine social goal, such as security.

More broadly, we might try to develop output indicators for science as measures of advancement of a series of social missions acknowledged to be science-intensive: e.g., industrial innovation, national defense posture, national health status, food production, environmental quality, energy independence, or capability in space missions. The major difficulty with this type of measure is that the success of R&D may be a necessary but far from sufficient condition for achieving favorable measures of performance in these areas, even if we accept the goals themselves uncritically. In most of the missions cited, science may not even be the most important factor. For example, in the case of industrial innovation, there is a vigorous debate as to whether the generally acknowledged lag in U.S. industrial innovation is due to a decline in the effectiveness and productivity of the national R&D effort, or whether it lies primarily in the steps that come after R&D and is more a function of failures in tax or regulatory policy or even a breakdown in our educational or managerial system.[8] In addition, progress in the achievement of these social missions is probably more a function of the cumulative stock of knowledge than of the current rate of acquisition of knowledge at the frontiers of either science or technology. In consequence, indicators of mission success have a built-in time lag that makes them insensitive to the current health of the technical enterprise. The same observation applies, of course, to some of the more intrinsic indicators such as Nobel prizes, which appear to depend more on the cumulative result of past scientific activities than on current rates of progress.

In considering the health of science it is increasingly acknowledged that R&D is part of a much larger process of industrial and societal innovation, not all of which is science-intensive or even science-dependent. In recent writings on science policy, such as those of the Organisation for Economic Co-operation and Development (OECD), stress has been laid on the idea that national and international science policies need to be much more closely integrated with economic policies as well as with social policies for specific sectors such as the environment, health, or education.[9] But ways of reflecting this integration—either in setting R&D priorities or in developing indicators of progress—have been slow to evolve, in part because of the difficulty in determining the degree to which shortfalls in performance in various sectors are the result of shortfalls in R&D performance or investment, or arise from other factors.

A second problem in developing extrinsic indicators arises from the fact that the most appropriate measure of the success of today's innovative efforts is not the solution of today's problems, but its success in anticipating and dealing with the problems that will be paramount down the road, when today's research results become available for social or economic implementation. Will the science we are generating today be the science wanted and needed to attack the salient problems two decades hence? This requirement to anticipate future societal targets makes the function of technology assessment increasingly important both in setting scientific priorities and in creating appropriate organizational settings for the management of science and the allocation of resources.[10] We must constantly avoid the temptation to organize innovative effort largely or exclusively around contemporary concerns. At the same time, while technology assessment directs our attention to significant areas of ignorance, and thus provides guidance for research directions, it is nevertheless an incomplete guide. Many future problems will be generated from outside science and technology, and are therefore difficult for technology assessment to identify. In particular, it is difficult to foresee changes in social expectations and sensitivities. For example, society is making demands on science to demonstrate or achieve a degree of risk-reduction for a variety of economic activities that would have been difficult to anticipate fifteen or twenty years ago, even if we had known as much about the objective threats posed by these activities as we do now. In general, society now demands a much higher burden of proof with respect to the safety and environmental suitability of deploying new technology than was true twenty years ago. One could say this has come out partly as a consequence of many well-publicized examples of past lack of foresight in the use of a variety of new technologies, but public sensitivity has nevertheless increased in terms of the absolute standards applied.

Problem Anticipation—Some Illustrative Examples

In this section I will analyze four examples of problems which society faces as a result of past technological activities, and whose mitigation might have been assisted if more anticipatory research had been conducted. What might we learn from these examples—the disposal of toxic chemicals, the management of radioactive wastes, reducing the toll of automobile accidents, and the environmental aspects of the cancer problem—that could be applied to setting research priorities today in order to improve our ability to deal with tomorrow's problems?

Toxic chemicals

The chemical revolution of the last thirty years represents one of the great positive accomplishments of technology, but it has left in its wake a large number of waste dumps whose chemical content and potential health effects on people are largely unknown. In the words of one observer, "we know enough to be apprehensive, but we don't know enough to really assess the situation as thoroughly as we would like."[11] Chemicals are disposed of, usually in landfills on the premises of the manufacturer, with little knowledge of their ultimate fate in the environment—their possible chemical transformation to other more or less toxic substances, their migration through the soil to contaminate groundwater, the likelihood of people being exposed to them, and the toxic effects to humans of possible levels of exposure. We do not know enough today to estimate the relative contribution of toxic waste dumps to human health risks in comparison with other potential sources of chemical exposure.[12] Nor are we in a position to estimate the cost-effectiveness of various techniques either for safely disposing of chemical wastes in the future or for acceptably and permanently cleaning up past waste sites. Many years of research will be required to settle these issues satisfactorily, and "a lot of technical people have come to realize that this is an issue that has gotten too little attention for too long."[13]

The question is, could we have done better? Why did we not begin to accumulate the basic knowledge and techniques fifteen or twenty years ago which would have enabled us to plan and execute the necessary applied R&D for assessment and disposal today? Whose responsibility should it have been to anticipate these questions and direct research efforts accordingly? Possibly only the industries involved could have had sufficient detailed awareness of the potential problems to have developed a relevant program, but industry in the 1940s and 1950s operated within a far different climate of public attitudes and expectations toward their responsibility to society than is the case today. The burden of proof was more on those who wanted to show cause for concern. The law did not, as it does today, make industry liable for the environmental and health effects of its effluents. The set of societal expectations embodied in the Toxic Substances Control Act of 1976, in the Resource Conservation and Recovery Act of 1977, and in the superfund legislation[14] of the late 1970s, has far outrun the existing capabilities of science and technology, yet at the time the problem was developing there was nothing in the political or legal framework to motivate or attract support for research on the fate or effects of toxic chemicals.

However, to say that the general climate of opinion in the 1940s and 1950s would not have supported the relevant research cannot be a sufficient answer. Some argue that it was the responsibility of the scientific community to identify the upcoming problems and mobilize political support for the necessary research effort. We might ask if present mechanisms and institutions for technology assessment, had they been in place in the 1950s and 1960s, would have focused sufficient attention on the problem to attract financial resources and fundamental scientific interest to this area of research? Or are the associated research problems so specific to particular waste sites and to the effluents of specific manufacturing processes that it would have been impossible to define generic scientific issues that could have been attacked by the general scientific community? This is, of course, a hypothetical question that is difficult to answer.

Many of the necessary analytical techniques were not available; but, during the period in question, analytical chemistry was a rather unfashionable and backward subject in university chemistry departments. There, priorities for research were governed largely by who got appointed to what "professorial chairs," and this in turn was determined by peer judgments as to where the most exciting scientific opportunities lay. In this competition, analytical chemistry took a back seat for several decades, compared with physical chemistry, synthetic organic chemistry, biochemistry, and chemical physics.[15] The emergence of new and more sensitive analytical techniques in the 1970s was a product not primarily of developments within chemistry but of the application of new tools imported from physics by chemists working in the fashionable chemical subdisciplines. It is open to argument whether the pace of developments in analytical chemistry could have been forced by greater attention to analytical chemistry as such, especially if it had come at the expense of support for the neighboring disciplines that ultimately provided some of the most important advances in analytical sensitivity.

The major potential impact of the disposal of toxic chemicals is on groundwater which may later come into contact with humans, primarily through withdrawal for drinking water supplies. Today, the United States has nearly 200,000 industrial landfills, lagoons, and waste impoundments for disposal of chemical wastes, all of which ultimately communicate to groundwater to an unknown extent. In addition, there are some 400,000 waste injection wells, with an estimated 5,000 new wells added each year. Perhaps 95% of these sites have, at present, no system of groundwater monitoring.[16] The use of land disposal for chemical wastes may have been encouraged in a variety of ways by the concentration of recent public concern and regulation on waste disposal in surface waters, the oceans, and the air. In addition, most land disposal takes place on property owned by the disposer, which until recently was not viewed as part of the "commons" that received most of the environmental movement's attention.

There was also no systematic research to compare different disposal media in relation to their potential human impact. Here again, one might argue that the subject of groundwater movement and exchange with soil media was a very "messy" research topic, one which did not lend itself to elegant and widely applicable generalizations, and hence was unfashionable among geologists. Moreover, the characteristics of each disposal site tend to be unique. Individual analyses of each of the many thousands of sites would have constituted a task absorbing very large resources and manpower compared to what was available for fields of geology with greater conceptual challenge.

Should solving the question of safe, long-term storage of toxic wastes be an industrial responsibility, like the development of new techniques for petroleum exploration, or should it be viewed as a challenge to the broader geological community? Unlike the situation for oil exploration, the economic rewards for new information about groundwater were not apparent in the 1950s and 1960s and, from the point of view of the chemical companies, might even have been negative, adding greatly to their costs. In other words, neither the economic reward system of the marketplace, nor the prestige and recognition system of the scientific community, nor the political reward system of public officials and legislators, generated the incentives to invest resources and effort in this important field. Although the environmental movement was beginning, at the time, to generate political rewards for attention to the environment, the focus, as we have seen, was on the recognized "commons"—air, surface waters, and the oceans.

Radioactive wastes
Here we have a rather different case. Unlike the situation with chemical wastes, the potential problem was recognized by scientists and engineers long before the associated technology became a commercial reality. One could argue, in fact, that the main elements of the problem, and the kind of research necessary to cope with it, were well understood almost from the beginning of the nuclear program. Most technical experts have viewed problems of radioactive waste disposal as "solved in principle" for the last two decades, yet only in the last half dozen years have significant manpower and resources been devoted to the problem, especially when viewed against the background of the entire public and private investment in nuclear technology, both military and civilian. From the beginning, popular and expert appraisal of the magnitude and importance of the problem differed considerably.[17] The experts, believing that a solution in principle existed, felt that practical implementation could be safely postponed until after a significant industry existed, especially since the longer the wastes "cooled" in temporary storage, the easier would be the final disposal, due to lower specific activity and internal heat generation. The experts totally underestimated the public's apprehension about the dangers of radioactivity. Its original association with nuclear war, and its insidious, invisible character, made it much more fearsome to the public than to the experts who had worked with it on a familiar basis for many years.

Although nuclear development, both military and civilian, attracted some of the most talented manpower available in the postwar period, the subject of waste management remained a scientific backwater with very low prestige even within the nuclear community. Of the prestigious Fermi awards, and the more numerous E.O. Lawrence awards, given out by the Atomic Energy Commission to recognize important contributions to the development of nuclear energy, only one went to an individual for work even remotely related to radioactive waste management.[18] Research on the problem was left largely to chemical engineers, and the relevant knowledge of geologists and materials scientists was not tapped until quite late in the game. This situation must be contrasted with other aspects of nuclear development, where the contributions from all relevant disciplines and engineering experience were eagerly sought. The importance of site-specific geological investigations to characterize potential waste repositories was seriously underestimated. This, in turn, led to an underestimate of the sheer hard work required to go from the individual technical elements of a solution in principle to the engineering of a specific waste repository, especially one with sufficient assurance of long-term integrity to satisfy skeptical critics and the public.[19]

Much of the geological and hydrological science required for radioactive waste management has potential applicability to toxic chemical wastes as well, although different depths and geological structures are involved. Had research and development for radioactive waste management, both high-level and low-level, been taken more seriously in the 1950s and 1960s, we would today be in a much better position to deal with the (probably) more difficult and ubiquitous chemical waste problem. Indeed, research on the dispersion of radioactivity in air, water, and soil, and in the food web, limited as it may have been from the standpoint of waste management, is still much more extensive than for chemical contaminants, and has probably advanced our knowledge farther than where it would be if nuclear technology had not existed. Even the weapons testing fallout gives invaluable tracers with which to study ocean circulation and to understand the oceanic sink for CO_2 which results from the burning of fossil fuels.[20]

Nevertheless, the scientific community's failure to give sufficient priority to radioactive waste management at a time when it was in a strong position to influence priorities must be counted as a failure of responsibility. Chemical wastes might legitimately have been considered as an industrial responsibility—there being little governmental support for chemical technology—but the general scientific community's much more extensive involvement in nuclear energy, and its strategic advisory role in government research allocations, make the failure less excusable. As in the chemical waste case, however,

Auto safety

Morison has suggested that our ignorance about accidents, particularly automobile accidents, represents another example of the failure of the scientific community to assign adequate priority to a continuing problem that is a direct outgrowth of an important technology.[21] Accidents are now the principal cause of death of persons under 35, and auto accidents comprise more than 50% of this toll.[22] Yet only a handful of U.S. research centers do significant work on the causes of auto fatalities and ways of reducing them. Such public measures as have been taken to mitigate the appalling toll are based largely on conventional wisdom rather than the results of empirical research. Neither the industry, not the technical community, nor the government has seen the "cure" of the auto fatality "disease" as a responsibility comparable in priority or challenge to heart disease, cancer, or stroke—all of which are the subject of massive, politically popular research efforts attracting some of the best talent in the biomedical profession. Research on auto accidents was—and still is to a somewhat lesser extent—a backwater of science which received little recognition and did not generally attract the best talent. The auto industry persuaded itself and the public that auto fatalities were the fault of "the damn fool drivers" and nothing much could be done about them. With little evidence to support its value, universal driver training has been the only panacea of public policy adopted on a large scale.[23] When the small amount of ongoing research began to identify vehicle design as a significant parameter affecting the severity of injury and the probability of death in an accident, Ralph Nader seized on this evidence to launch a crusade to force the auto companies to take injury mitigation seriously in their designs. New legislation, and a new Auto Safety Administration with legal power to set safety performance standards for automobiles, was the result.[24] However, in this case, as in the case of toxic chemicals and chemical waste dumps, legislation stimulated research rather than vice versa. Although one can argue that research provided some ammunition for the legislative campaign, it still seems true that social expectations embodied in current legislation and regulation have moved far ahead of the research.

The auto accident example differs from the preceding cases of chemical and radioactive wastes in two important respects. First, it constitutes much more of an actual than a potential threat to human health and life. The number of people killed and maimed in a year exceeds 100,000, whereas deaths and injuries from toxic chemicals are isolated and the number of proven cases is small. Injuries from man-made radioactivity are practically non-existent. Second, mitigation of the auto toll is probably much more dependent on the feasibility of changing human behavior on a large scale, and less on technical solutions. In other words, the potential role of science and research in dealing with auto accidents is less obvious than in the other two examples. The implementation of already known measures—such as mandatory seatbelt use, removal of alcoholic and psychotic drivers from the roads, and changes in highway design—lags behind the research that is already available. This is partly because the enforcement of auto safety measures collides with notions of individual rights and entitlements. The definition of research "needs" is made uncertain and problematical by doubts and controversies over the feasibility of implementing the results, even after they become available. In this situation, the mismatch between technical opportunities and apparent social needs may be a significant factor in explaining the research community's failure to enter into the problem wholeheartedly. The interdisciplinary nature of the problem, the interaction of technological and behavioral factors, and the high perceived public and private benefits of personal mobility all combine to make auto safety an unusually messy problem from the standpoint of formulating a consensus on suitable research approaches. Thus it is not clear whether better anticipatory research or technology assessment fifteen or twenty years ago would have made much difference in the actual status of the problem today.

Cancer and the environment
Another example of misplaced scientific priorities cited by Morison[25] is the comparative neglect of environmental factors in the causation of cancer, and what he suggests may have been the excessive funding of research on the cellular aspects of cancer. Some experts interpret epidemiological studies over the last two decades to mean that be-

Indicators and Priorities

tween 60% and 90% of all human cancers are of environmental origin and thus theoretically preventable through control of what gets into the environment and what materials people come in contact with.[26] The meaning of the term "environment" is vague, and the conclusion concerning environmental influences is often misinterpreted to suggest that most cancers are caused by air, water, and soil pollution arising from industrial and agricultural activities. In fact, the term "encompasses anything that interacts with humans, including substances eaten, drunk, and smoked, natural and medical radiation, workplace exposures, drugs, aspects of sexual behavior, and substances present in air, water, and soil."[27] Many of the sources may be natural rather than man-made. Probably the largest single environmental factor is diet, and "nutritional habits that affect hormonal and metabolic balances are believed more important than additives and contaminants."[28] However, ignorance of the role of specific environmental factors is almost total, except in a few cases such as smoking, occupational exposure to asbestos, and ionizing radiation (primarily from natural background rather than man-made). Evidence for the importance of environmental factors is quite indirect. It comes from geographical differences in the incidence of malignancies, combined with the fact that these differences tend to change when a population moves to a new location.[29] Is this an example of a situation in which the scientific community failed to anticipate future problems, despite obvious evidence, while pursuing interests dictated by its assessment of scientific opportunities rather than societal needs? To some extent the answer is yes, although it is difficult not to be biased by the wisdom of hindsight.

The issue is somewhat more starkly illustrated by the fact that the National Cancer Institute (NCI) established its SEER (Surveillance, Epidemiology, and End Results) program only in 1973. This is a continuous cancer incidence reporting program, the first of this degree of thoroughness in the United States. Based on a sample of only 10% of the population, it is currently unrepresentative of the actual demographic make-up of the country, and has been in existence too short a time to permit detection of the influence of environmental factors, which often have latency periods of 20 to 40 years.[30] The cost of the present data collection program is about $10 million annually, only 0.1% of an NCI budget of $1 billion.[31] In a recent report on technologies for determining environmental cancer risks, the Office of Technology Assessment proposes 11 possible governmental policy initiatives in this area. Four of these proposals deal with increased efforts to gather data on the incidence and distribution of cancer as a function of selected population characteristics and history. Three deal with programs to improve methods of testing chemicals for carcinogenicity, and the implementation of these methods on a larger scale than at present.[32] All of the proposals contemplate massive data collection and testing efforts which are routine, centrally organized, and highly systematic and prescribed—hardly an activity likely to be attractive to creative scientists. Several of the recommended programs could easily have been mounted twenty years ago. Because of the significance of cumulative, longitudinal information, the value of such programs increases in proportion to their duration. It could be argued that our knowledge about cancer would be far more advanced today if this approach had been initiated earlier. But it would also be legitimate to ask what would have been lost to science if the National Cancer Institute had spent, say, $50 million a year on the SEER program and 0.4% of its total budget less on the mix of other research programs. As early as 1964, the Wooldridge Study evaluation of the National Institutes of Health could be interpreted as advocating that such systematic data collection and testing was a responsibility of the in-house programs of the Institutes which they were not adequately fulfilling.[33] But the political climate of the times was against this opinion—which, in any case, was couched in such terms as to appear too self-serving for the interests of the extramural university programs.

Had something like the Office of Technology Assessment existed twenty years ago, and had it produced the type of assessment of the environmental cancer problem that it was able to provide in 1981, with broad substantive input, advice, and review from a diverse outside technical community, there could have been sufficient influence on research priorities to affect what we know about cancer today. It is less certain, however, that such changes in our knowledge would have affected our ability to prevent or cure cancer. As in the case of auto accidents, we may question whether the implied behavioral and lifestyle changes suggested by this type of research could really have been implemented. The lessons from our experience with research on smoking and cancer (which has been pursued vigorously during the last twenty years) suggest that the risks must be sufficiently large, relative to the perceived benefits of many habits and lifestyles, to cause people to respond to additional contradictory information. Only in the (probably small) minority of cases where the harm can be traced to artificial contaminants placed in the environment by industrial activities does it appear likely that large changes can be effected. This is, of course, speculative. It could also be argued that research on methods for effectively communicating research findings and influencing the behavior of large numbers of people could also have helped. Yet, the probability of success from large-scale implementation of therapies derived from cellular-type research on cancer seems much higher. Judging by the smoking experience, the establishment of correlations between environmental variables and cancer seems inadequate evidence without deeper scientific insights into the actual causal mechanisms. Here, the molecular and cellular approach for studying the mechanisms of carcinogenesis which underlie the epidemiological data and empirical tests seems more promising, even simply in terms of convincing the public to take the results seriously.

Conclusions from the examples
In each of the examples reviewed above, the existing research planning systems have resulted in underemphasis on systematic and well-planned monitoring, data collection, and systematic empirical testing or environmental characterization. The kind of research involved is low-grade, often routine, and scientifically not very challenging. It is not likely to attract the interest or attention of independent investigators, and it requires much greater managerial effort and organization than traditional investigator-originated research, even in normal "big science" or in clinical investigation. Where such data collection is proposed within the peer-reviewed project grant system it is likely to get a low rating for scientific quality, and not to fare well in competition for limited resources. It is more appropriate for government in-house research, or for research contracted to specialized organizations offering continuity and a degree of cumulative learning.

Any monitoring or data collection system should be subject to a continuing process of review, assessment, and possible revision, by outside advisory groups including scientists, laymen, and policymakers. The review should be much broader in its disciplinary representation than the customary peer review panel or research advisory committee to an agency.

We need to learn how to use technology assessment as a tool for research planning, as a means for identifying gaps in knowledge and for influencing (although not determining) directions for future research and data collection. Periodic updating of a diversity of technology assessments in selected problem areas could serve as a supplementary type of science indicator, marking progress toward certain social goals of high technical content and guiding applied research and certain kinds of basic research. The recent OTA assessments of technologies for determining cancer risks, and of Recombinant DNA technologies,[34] might well be used as pilot exercises for such periodically updated technology assessments used as indicators.

The Future

If in the past we have fallen short in anticipating the key future problems and directing national research efforts toward them, what are the prospects for improving our performance? One point needs to be emphasized. The expertise required to identify and formulate sociotechnical goals, and to relate research objectives to social objectives, is different from the expertise required to assess the opportunities and prospects for advances in a scientific discipline or a technological development. On the other hand, in a technological, interdependent world, conventional wisdom and popular political preferences are also not a sufficient basis for establishing goals (even for applied research) or for assessing progress toward those goals. The notion that society and the political system should set broad social goals, leaving the technocrats and experts to determine the tactics for reaching these goals is naive and unworkable. Means and ends affect each other too intimately for such a separation between goals and tactics to be viable. Setting priorities, even for basic research, must involve a many-sided and continuing interchange among scientists from several disciplines, intelligent and concerned laymen, and specialists in the management and planning of operations in specific sectors such as the running of cities, the maintenance and enhancement of food production, health care delivery, or environmental protection. Most of the existing literature on the setting of research priorities deals with the comparison of large scientific or technological projects,[35] or the comparison among disciplines within basic research, rather than with the planning of research in relation to explicit social, economic, and political goals.[36] The effort has been to find a rationale for rating the relative importance of different scientific disciplines in some general relationship to what the raters consider to be the society's current goals.[37] To the extent that social goals are introduced more broadly into the discussion of research priorities, they tend to be taken as givens, whereas in fact the essence of research goal setting should lie in the tough intellectual problem of mutually adjusting social goals and research goals to a viable set of objectives which can no longer be viewed as purely social or purely technical. The key to the task of setting priorities and measuring progress in this sense is the proper structuring of the diverse interactions and dialogues among experts and other publics so as to progress toward a consensus that has a reasonable degree of stability over time, while retaining a flexible framework within which the technicians and the scientists can easily, continuously, and rapidly adapt their tactics to constantly changing information and new analytical insights. This is a large order, and we can only dimly envision how to tackle it. As we will explain more fully in the next section, the task is further complicated by the fact that it cannot be formulated in purely national terms, even within the largest national states.

In order that research be better aimed at future problems rather than today's problems, an essential element in setting research priorities and measuring accomplishments is some kind of projection of the future, both social and technical. Such a projection is not really a forecast, but a hypothesis or assumption usable for planning. It need not be detailed, and it may cover a range of plausible possibilities or "scenarios" rather than only one. A more refined approach would be to project the future based on an extrapolation of contemporary trends and then to estimate the change in projections if various scientific or technological developments were successful and were actually implemented in investment decisions either through the market or via public policy or some combination of the two. In recent years there have been a number of attempts of this kind. One which explicitly includes the step of relating changes in projections to possible R&D strategies has been published by the International Energy Agency (IEA) of the OECD under the title, "A Group Strategy for Energy Research Development and Demonstration."[38] It is worth quoting the announced objectives of this study in full, because it illustrates what I have in mind more generally:

1. To provide an assessment of the likely relative importance of individual technologies for the IEA nations as a group;
2. To obtain estimates and from these develop targets for the energy impact to be achieved by new and improved technologies during the latter part of the century;
3. To provide a tool for:
—developing and assessing national R&D policies and plans
—achieving the most effective relationship between the R&D activities of the member countries; and
4. To identify non-technology policy issues which can affect the ability of new technologies to contribute to energy requirements.[39]

The philosophy of such an approach is easier to state in principle than to carry out in practice. The energy sector is strongly coupled with many other aspects of society, and its projected evolution will depend on many assumptions about political, economic, social, and demographic developments. This will, in fact, be true for almost any other sector chosen for a similar analysis. Projections based on current trends that can already be easily discerned are inevitably "surprise-free" and hence tend to overlook those "clouds no bigger than a man's hand" on the horizon which could turn out to be dominant influences in twenty years. Typical examples are wars, revolutions, unforeseen demographic shifts due to migration, political ineptitude, unanticipated secular shifts in public opinion due to generational changes, technological surprises, a sudden cutback in world petroleum production or exports. Moreover, in order to reduce the complexity of the problem, there is a strong propensity to minimize the interactions between problem areas—between energy and environmental goals, between energy and the international financial system, between energy policy and political alienation among youth—to cite just a few examples.

In view of these many sources of uncertainty, some argue that it is better to cover all plausible bets in R&D, and thus be guided more by technological opportunity than by the uncertain projected socioeconomic impact of various options. Others argue for letting the market—perhaps modified to approximate proper internalization of externality costs—decide. Still others would argue that the lead times in some types of energy development are so long that market incentives are insufficient to induce private entrepreneurs to take the initial steps in a long sequence of undertakings fraught with both technical and market risks at every stage. This last group would advocate a more active government role even in sectors where the ultimate delivery of goods and services is acknowledged to be a responsibility of the private sector. However, even if the projections of socioeconomic impact should not be the major determinant in R&D planning, it cannot be treated as irrelevant. For example, large public investments in R&D, where the ultimate socioeconomic impact appears to be small over a wide range of plausible assumptions about future conditions, should probably receive low priority. Conversely, developments with large potential impact over a wide variety of assumed futures demand close attention even if there are some circumstances in which their impact would be small. The greatest difficulties lie in the in-between areas where the possible impacts vary widely between assumptions of nearly equal plausibility. In deciding whether to invest in RD&D, the greatest differences of opinion will occur over the relative weight to be given to technical opportunity as opposed to projected socioeconomic impact.

Future energy supplies and demands have received by far the most attention among different sectors with respect to the determination of RD&D strategies. Both private and public agencies have published hundreds of projections in the years since the 1973 oil embargo and subsequent oil price increases.[40] A more ambitious approach, attempting to transcend the single-sector emphasis, is represented by the *Global 2000* report, a projection of world developments in the next twenty years.[41] Undertaken as a cooperative venture by many different U.S. government agencies, this effort was led by a task force jointly chaired by the Council on Environmental Quality (CEQ) and the Bureau of Oceans, Environmental, and Scientific Affairs (OES) in the U.S. Department of State. The report attempts to bring together the projection models used by many different agencies into a common framework of assumptions, thereby generating a single, internally consistent projection of many different aspects of the world's future—the rich-poor gap, population growth, world food production, economic growth, energy supply and demand, environmental deterioration. The consensus of reviews suggests that *Global 2000* was a noble and commendable but largely unsuccessful effort.[42] Disparate agency models were not reconciled; therefore, the assumptions underlying projections for different sectors such as food and energy were inconsistent with each other, and there was insufficient opportunity to adjust the assumptions of one model to take into account the output of another in an iterative manner. From the standpoint of this essay, a more fundamental difficulty was the failure to account for the impact of various possible scientific and technological developments on the projections. The assumption of "continuation of existing trends" did imply an allowance for continuation of past productivity growth rates in some sectors such as agriculture and manufacturing, but did not even consider what kind of research and development would be needed to insure this. Thus, the project missed an interesting opportunity to develop a framework within which useful public discussion of R&D priorities could have carried forward. Nevertheless, more of this kind of multisectoral analysis is needed to relate R&D more clearly to social goals.

If used as part of a system of science indicators, such projections should be updated regularly. At the same time, there should be systematic data collection that emphasizes monitoring parameters expected to indicate the impacts of various R&D efforts and their implementation. A long period of continuous data collection would be necessary before the indicators could significantly measure output from R&D or innovative activity; but, over extended periods of time, the data could provide meaningful feedback to research planning.

I am not suggesting that the social output indicators sketched above could ever wholly replace the more inward-looking science indicators in use today, or that setting research priorities on the basis of potential socioeconomic impact should entirely replace peer review and self-management by the scientific community. An opportunity-oriented strategy in a significant fraction of the technical enterprise must always remain essential to the healthy development of science, and thus ultimately to its *capacity* to contribute to social development. The social output indicators supplement rather than replace; they are intended to monitor society's *utilization* of science, and hence they measure not only the effectiveness of R&D but the effectiveness of all the stages of innovation beyond R&D which are necessary for its social or commercial implementation.

Many pieces of such data collection, monitoring, and system modeling are already in place or under development as a result of work started in the last twenty years or of new institutions created for related purposes. I have already mentioned the numerous energy models and projections and the *Global 2000* report. The system of linked modeling programs created by the International Institute for Applied Systems Analysis (IIASA) under the combined auspices of twelve nations on both sides of the Iron Curtain is another example.[43] The East-West Center in Honolulu has a somewhat similar purpose.[44] Several of the special data collection and monitoring programs set up under the United Nations Environmental Program (e.g., the Global Environmental System) and under UNESCO (e.g., the Man and the Biosphere Program) contain elements of what I am suggesting. Several of the programs being undertaken by the International Federation of Institutes for Advanced Study (IFIAS), particularly the so-called ABC Program (Analyzing Biospheric Changes), involve an attempt to deal more definitively with the interactions among several different problem areas, for example, the cultural response to biospheric changes.[45] These are only a representative sampling, not a comprehensive list, of such activities. However, none of these efforts sufficiently emphasizes feedback from monitoring and data collection and modeling to the global R&D agenda, or the potential impact of possible scientific and technological developments in modifying systemic changes in the future. For example, if there is widespread industry substitution of microbiological processes for chemical processes, what might be the implication for the control of toxic chemical emissions to the environment? Assuming the impact is favorable, what would be required to accelerate it? What should we look for? To what extent can genetic engineering help convert persistent toxic organic wastes to harmless substances with less environmental impact? How may the widespread application of information technologies reduce society's requirements for primary energy and materials without reducing the amenities now provided by the consumption of such resources? To what extent might it be possible in the application of sophisticated information technology to stress materials and energy saving rather than labor saving, and would the result be desirable for society? What are the prospects for substituting biological for chemical methods of pest and pathogen control in agriculture, and what would be the cost of such an effort? What are the prospects in the long term for the creation of industrial methods of food production which are less intensive in the use of land, water, and soil nutrients? What new knowledge and technology is required to increase the efficiency with which we use freshwater resources, and when would the benefits justify the expenditures of effort and resources? What are the prospects for compensating the climate modifications due to CO_2 and other trace gases in the atmosphere by deliberately engineering albedo changes over the earth's surface? To what extent will the increased mobilization of nitrogen and phosphorus resulting from human activity help to "lock up" the additional CO_2 in the atmosphere produced by fossil fuel combustion? In what ways could this process be deliberately "helped"? What would be the effect of the realization of any of the above technical achievements on the evolution of the global scenarios derived from the modeling effort?

International Considerations

Much of the literature on both science indicators and research priorities focuses on national science policies—primarily because virtually all public support for research is channeled through national budgets, even when it is contributing to or coordinated with international or multinational efforts.[46] Even the industrial R&D funds of multinational corporations are usually allocated and managed through a corporate headquarters located in the parent country. R&D is an activity of sovereign nations; thus, its planning and impact are considered largely within a national framework.

Yet the results of a large proportion of both public and private research belong to the international technical community, and can be applied or implemented by any country with the requisite minimum technical competence, irrespective of whether it was the originator of the work. Even most industrial proprietary information can be rather freely traded across national boundaries, and it is generally believed that the cost of purchasing such proprietary information is less than the cost of producing it in the first place. Although the results of much military-related R&D remain classified, their monopoly by the originating country appears to be quite short-lived.[47] In addition, most of the more generic types of military research are unclassified and rapidly disseminated through the technical literature because of their multi-purpose nature.

If one is concerned with the outputs of a country's R&D system rather than with its inputs, then national statistics and national planning are of dubious significance, especially in the longer term. Countries such as Japan, Germany, and Italy, which enjoyed unusually high economic growth rates during the early postwar period (at least up until the early 1970s), based a large proportion of their industrial growth on imported rather than indigenous technology. The United States, which had the strongest R&D programs throughout this period, had slower economic growth and an extremely high ratio of exports to imports of technological know-how.[48] In addition, the United States exported considerable technical knowledge through the training of foreign scientists and engineers, especially at the postdoctoral level.[49] A surprisingly high proportion of major industrial innovations originated in the United States and the United Kingdom but were first applied on a large commercial scale by other countries, especially Japan.[50] To some extent this was the result of an inevitable catch-up period after the destruction of World War II, and today the origination of new technology is much more evenly distributed relative to population than it was during much of the postwar era. However, the circulation of technology across national boundaries, and the speed of imitation and diffusion of new technologies are probably higher than ever, thus further reducing the significance of a purely national focus in science policy.

National science policies that fail to take full account of concomitant world developments, and that fail to consider the international division of labor in science and technology, will be increasingly wasteful and unproductive. The return on investment for imitative discoveries or inventions drops precipitously from that of first discoveries, unless compensated by a superior capacity for rapid exploitation. Furthermore, the nation which comes off second in the generation of an innovation might have received the same economic or social return from a smaller research investment combined with a system for scanning world R&D results which would permit rapid identification, purchase, and commercial pursuit of relevant developments made by others. Of course, in a world where all the leading industrial nations are not at roughly the same technical level, there is the danger of a "free rider" problem. Most countries wait to see what the pioneers do, in order to benefit from their discoveries and, it is hoped, avoid many mistakes and blind alleys. If all countries followed this strategy, however, then innovation and research would drop sharply on a world basis; but there is as yet no indication that this is happening.

Another problem with planning R&D on a national basis is that the benefits of research are increasingly transnational, and yet these benefits are not considered in setting national priorities. Indeed, research with transnational benefits tends generally to be undersupported in the present system.[51] This is a particularly acute problem for lesser developed countries (and for the rich-poor gap). It also results in lagging progress in such areas as environmental protection and conservation of biospheric resources, where the benefits do not accrue exclusively to individual countries. It also applies to most fundamental research.

The difficulty of setting R&D priorities and assessing national performance is greatly exacerbated for small countries that can only afford to stay abreast of the advancing world frontier of science and technology in a small number of carefully selected fields. In which fields should such a nation invest to remain marginally ahead? And in which fields should it merely maintain enough activity to keep a "watching brief," being ready to imitate and improve upon breakthroughs when they occur elsewhere? What is the minimum level of effort required for a field of science to be able to take adequate advantage of discoveries or new developments originated elsewhere? In other research areas, such as those relating to environmental quality or health and safety, or to the provision of goods and services that are only sold or provided locally, the decision may be even more complicated. As suggested by Raymond Vernon, perhaps science indicators should be designed to measure the rate of adoption or adaptation of commercial technology, or the rate of incorporation into public policy and regulation of world research results bearing on health, safety, or environmental quality,[52] rather than the origination of new research results. Such measures would be extremely difficult to design in practice, but might be worth experimenting with in exemplary cases as a first step toward the eventual design of a much more comprehensive system of indicators.

For the large industrial countries, the problems of priority setting are also difficult because of the responsibility that such countries currently bear for the system as a whole. For example, the United States accounts for roughly 30% of the world R&D investment, and perhaps as much as 50% on the average of the world output of scientific papers.[53] Any change in these U.S. indicators would have significant repercussions for the growth of the world pool of technical knowledge, and hence for global capacity to cope with societal problems of the future. The United States posture vis-à-vis the world knowledge pool may have aspects analogous to its situation vis-à-vis international security. United States expenditures on both military R&D and military procurement and readiness are out of proportion to those of many other industrial countries to which it is allied or which partly depend on this spending for their protection. Is the

U.S. carrying more than its "fair share" of military RD&D as well as of general purpose R&D (such as biomedical) to the detriment of its competitive position in international trade and domestic social welfare? Because of this, is there a *de facto* inverse relationship between traditional input science indicators and other potential measures of the benefits of U.S. scientific and technological investments? Or, on the other hand, is the enormous U.S. investment of technical resources in economically sterile military R&D depriving the world of a desperately needed capacity to deal with the problems of population, resources, energy, environment, and economic development that it will face in the next twenty to thirty years? One can only raise these questions without answering them, but they serve to illustrate the limitations inherent in discussing R&D priorities and science indicators on a purely national basis.

Public versus Private Sector

In the setting of priorities for publicly supported research and the assessment of the "health of science," privately and publicly supported research and innovative activity must be considered together. The question of the public and private sectors' appropriate relative roles in R&D has been brought to center stage in science policy by the Reagan administration's initiatives to return many government R&D activities, particularly in energy, to the private sector.[54] It seems evident, in fact, that we are at the beginning of a new national debate on this question, one that is likely to draw attention for the next decade. Although probably no one seriously advocates withdrawing government from the support of R&D, the burden of proof has clearly shifted onto those who advocate the expansion of government R&D responsibility into new areas, and those who defend government support in the areas for which the government has only recently taken on responsibility (e.g., energy conservation and fossil fuel R&D). Only in areas where the market indisputably fails to channel adequate resources into R&D of widely recognized social value will the advocates of a strong government role be likely to prevail. The key argument is that decentralized decision-making about the allocation of R&D effort (such as derived from a competitive market) is more likely to filter and select information germane to choosing the most socially productive R&D than is any central planning system. The market, it is argued, is a more efficient information processing device than even the most sophisticated and comprehensive system of centralized planning and bureaucratic decisionmaking.[55]

Wide consensus appears to exist among even the most dedicated free-market advocates that, with market incentives alone, the private sector will underinvest in fundamental research, and that, therefore, government has a responsibility in this area. The same argument is accepted with considerably less unanimity in the international context, where it is likely that individual countries will underinvest in fundamental research of greater potential benefit to the world as a whole than to any single country. There is wide, but also less unanimous, agreement that government has legitimate responsibility for R&D that supports the assessment of health, safety, and environmental effects of emerging technologies because the market provides insufficient incentives for the generators of new technology to invest in this area.

On the other hand, many people maintain that externality and appropriability arguments like the above are much too easily abused. Almost *all* R&D has *some* benefits that cannot be captured through the market revenue from the resulting goods and services. These critics would insist, however, that to the degree for which this is so, it would be more efficient to recognize the fact through tax credits and other devices that leave the choice of actual R&D projects and strategies to decentralized market-like decisions. Similarly, many of the social costs of emerging industrial technologies could in principle be internalized through such devices as product liability and strict liability enforceable in the courts for compensation of victims of pollution and other negative externalities resulting from industrial activities. Thus, it is argued that direct government support of R&D in which choices are made centrally should be regarded as the mechanism of last resort for compensating for the failure of the R&D market.

Although it is obvious that the planning and evaluation of publicly supported R&D should be carried out in the light of the actions and plans of the private sector, this is not so easy in practice because private sector investments are frequently determined by what the private sector expects to acquire from the public sector. Thus it is hard to decide when a proposed public sector R&D program will simply displace something the private sector would have otherwise undertaken on its own. This is especially true if one assumes that the resources not used for this purpose by the public sector would otherwise have been available to the private sector for its discretionary use. The issue is whether the extra net social benefits derived from a public sector R&D program (taking into account that it may be less efficient) would exceed the sum of private and indirect social benefits from whatever alternative would have been chosen by the various private sector actors, including, of course, non-R&D alternatives. An additional advantage of public sector support for R&D is that the results tend to be communicated more widely, and the advantages for technical progress of open communication and sharing of information may far outweigh the competitive incentives arising from more exclusive reliance on the market. On the other hand, although one might be able to state with confidence for any given proposed public sector program that a wider set of interests would be considered than in the alternative private sector action, it might still be true that taxing a portion of the extra private sector returns from the alternative private sector activity would yield greater social benefit in the long run. It is this sort of consideration that makes R&D priority determination for the public sector an inherently speculative enterprise, and one in which it is virtually impossible to determine whether the right choice was made—even long after the fact.

As society becomes more complex and interdependent, and less inclined to tolerate the penalties to which losers are subject in an ideal competitive market, the external or third-party repercussions of all transactions within the society tend to increase in relative importance. Thus, the solution of one set of problems will have ever-increasing repercussions for the character and severity of other problems which would have appeared to be unrelated in a first approximation. To the extent that this is the case, one could argue that publicly financed R&D (or at least one important component of it) ought to grow more rapidly than private R&D in order to enable society in a collective sense to identify and deal adequately with the new problems generated as an ultimate consequence of private R&D. Whether this is a sound argument or only a plausible rationalization of a trend potentially beneficial to the scientific community is still a matter for speculation.

Conclusions

The discussion of research priorities and progress indicators is much more advanced in connection with fundamental research and the intrinsic characteristics of the scientific and technical enterprise than it is in relating research progress to non-scientific societal goals. This is, in part, because the goals themselves are insufficiently formulated for easy translation into technical objectives, and because relating technical and social goals is actually an ill-developed art that has hitherto been based largely on conventional wisdom and political clichés.

Several case examples from the past suggest that existing systems of research planning tend to undervalue systematic and well-planned monitoring, data collection, empirical testing, and localized environmental characterization. Such work is often dull and routine and absorbs large financial resources while not engaging the highest level of scientific talent. Yet the cumulative benefits of such work in terms of the capacity to anticipate future problems can be as important as more original and novel research. This is a problem that science policy has not effectively solved in any country.

In the last decade, considerable progress has been made in technological assessment, and in the construction of mathematical simulation models that can help forecast the consequences of continuation of present technical and social trends. However, much less progress has been made in utilizing such models to test the sensitivity of the projections of the future to a variety of possible technical achievements and thereby guide the allocation of resources to applied R&D. This has perhaps been best effected in the field of energy, but even here the practical guidance to public R&D investments has been minimal, in part because of inadequate follow-through and in part because of countervailing political forces. Relating R&D planning to technological and societal forecasting is in an even more rudimentary state than the forecasting techniques themselves.

Because of the increasingly rapid circulation of information derived from national R&D, especially among the advanced countries, the present system of largely national planning and budgeting for R&D is obsolete and inefficient. There is a need to develop mechanisms and institutions for negotiating R&D priorities on a multinational basis, making the results of such negotiation effective in the constituent national political processes, and monitoring progress in a similar multinational framework. This will not be easy, but it is urgent that it be attempted.

In the absence of a well-developed art and science for relating technical to social and economic goals, we are forced to fall back on the identification of technical opportunities as the main basis for allocating resources. On the whole we are better at identifying the potential benefits of specific technical advances than we are at identifying the specific technical advances required to deal with what are usually poorly defined societal problems. This is, perhaps, not as bad a strategy as it might at first appear; it is probably superior to reliance on conventional wisdom to relate technical and social goals. But, in a time of increasing resource stringency, we need something better.

Notes

1. Albert Rees, chairman, Panel to Review Productivity Statistics, *Measurement and Interpretation of Productivity Statistics* (Washington, DC: National Academy of Sciences, 1979), especially Chapters 5, 6, and 9.

2. Harvey Brooks, "Science Indicators and Science Policy," 2 *Scientometrics* 5–6 (1980): 331–337.

3. Charles Fried, "The University as Church and Party," *Bulletin of the American Academy of Arts and Sciences* 31 (December 1977): 38.

4. Robert S. Morison, "Needs, Leads, and Indicators," this volume.

5. Vannevar Bush, *et al.*, *Science, the Endless Frontier* (Washington, DC: U.S. Government Printing Office, 1945; reprinted July 1960 by the National Science Foundation).

6. Charles Fried, *op. cit.*, p. 38.

7. Harvey Brooks, "The Military Innovation System and the Qualitative Arms Race," in Franklin A. Long and George W. Rathjens, eds., *Arms, Defense Policy, and Arms Control* (New York: W.W. Norton & Company, 1976); Harvey Brooks, "Notes on Some Issues on Technology and National Defense," *Daedalus* (Winter 1981): 129–136.

8. Harvey Brooks, "Social and Technical Innovation," to be published in Sven B. Lundstedt and E. William Colglazier, Jr., eds., *Managing Innovation: The Social Dimensions of Creativity, Invention and Technology* (Elmsford, NY: Pergamon Press, 1982).

9. Organisation for Economic Co-operation and Development (OECD), *Technical Change and Economic Policy—Science and Technology in the New Economic and Social Context* (Paris: OECD, 1980); also see *Science, Growth, and Society: A New Perspective* (Paris: OECD, 1971).

10. *Technology: Processes of Assessment and Choice*, Report of the National Academy of Sciences to the Committee on Science and Astronautics, U.S. House of Representatives (Washington, DC: National Academy of Sciences, 1969).

11. "Probing Societal Risks," interview with William L. Lowrance, 59 *Chemical and Engineering News* 27 (6 July 1981): 13–20.

12. *Ibid.*

13. *Ibid.*

14. Toxic Substances Control Act of 1977, PL 94-469; Resource Conservation and Recovery Act of 1977, PL 94-580; and the Comprehensive Environmental Response Compensation and Liability Act of 1980, PL 96-510, 11 December 1980.

15. Compare, for example, the small amount of attention given to this subject in Frank H. Westheimer, *et al.*, *Chemistry: Opportunities and Needs* (Washington, DC: National Academy of Sciences, 1965).

16. "Groundwater Supplies: Are They Imperiled?," *Conservation Foundation Newsletter* (Washington, DC: June 1981).

17. Harvey Brooks, "The Public Concern in Radioactive Waste Management," *Proceedings of the International Symposium on the Management of Wastes from the LWR Fuel Cycle*, sponsored by the Energy and Research Development Administration, Conference 76-0701 (11–16 July 1976, Denver, CO).

18. Harvey Brooks, *ibid.*

19. *Energy in Transition 1985–2010*, Final Report of the Committee on Nuclear and Alternative Energy Systems, National Research Council (Washington, DC: National Academy of Sciences, 1979), Chapter 5, pp. 310–311.

20. W. Stumm, ed., *Global Chemical Cycles and their Alterations by Man*, a report on the Dahlem Workshop, Berlin, November 1976 (Berlin: Dahlem Konferenzem, 1977).

21. Robert S. Morison's remarks were made in discussion at the Harvard-MIT "Indicators of Quality in Science" seminar series, 1980.

22. Daniel P. Moynihan, Chairman, *Report of the Secretary's Advisory Committee on Traffic Safety* (Washington, DC: U.S. Department of Health, Education and Welfare, 29 February 1968), see especially Appendix C.

23. Moynihan, *ibid.*, Chapter V.

24. National Traffic and Motor Vehicle Safety Act, PL 89-563, 9 September 1966.

25. Robert S. Morison, *op. cit.* (Note 4).

26. Office of Technology Assessment (OTA), *Assessment of Technologies for Determining Cancer Risks from the Environment*, June 1981, p. 3.

27. *Ibid.*

28. *Ibid.*, p. 8.

29. Robert S. Morison, *op. cit.* (Note 4).

30. OTA, *op. cit.*, p. 20.

31. *Ibid.*

32. *Ibid.*, p. 19.

33. Office of Science and Technology, *Biomedical Research and its Administration: A Study of the National Institutes of Health.* (Washington, DC: The White House, U.S. Government Printing Office, 1965).

34. Office of Technology Assessment, *The Impact of Genetics: Applications to Microorganisms, Animals, and Plants,* 8 January 1981.

35. Alvin M. Weinberg, "Criteria for Scientific Choice II: The Two Cultures," 1 *Minerva* III (Autumn 1964): 3–14; reprinted in Edward Shils, ed., *Criteria for Scientific Development: Public Policy and National Goals* (Cambridge, MA: The MIT Press, 1968).

36. *Priorities for Research and Development, A Summary of Australian Work on the Identification of Priorities for R&D Bearing on National Objectives,* prepared for the Committee for Scientific and Technological Policy, OECD, Canberra, Australia, July 1981.

37. United Nations Educational, Scientific, and Cultural Organization (UNESCO), "Method for Priority Determination in Science and Technology," *Science Policy Studies and Documents* 40 (Paris: UNESCO, 1978).

38. International Energy Agency, *A Group Strategy for Energy Research Development and Demonstration* (Paris: OECD, 1980).

39. *Ibid.*, p. 9.

40. For a discussion, see *Energy in Transition 1985–2010,* Final Report of the Committee on Nuclear and Alternative Energy Systems, National Research Council (Washington, DC: National Academy of Sciences, 1979), Chapter 11.

41. Gerald O. Barney, *et al.*, Council on Environmental Quality and the Department of State, *Entering the Twenty-First Century: The Global 2000 Report to the President, Volume One* (Washington, DC: U.S. Government Printing Office, April 1980).

42. Alan S. Manne, "Energy Issues in Global 2000: Analysis or Advocacy?," internal memorandum to Office of Technology Assessment, U.S. Congress, 1981.

43. International Institute for Applied Systems Analysis, A-2361 Laxenburg, Austria. Annual reports and other publications are available from the Institute or from the U.S. National Technical Information Service in Springfield, Virginia.

44. East-West Center, East-West Culture Learning Institute, 1777 East-West Road, Honolulu, HI 96848. Annual reports are available from the Center.

45. International Federation of Institutes for Advanced Study, *Update: Programmes—Publications—Events 1980/81* (Solna, Sweden: International Federation of Institutes for Advanced Study, 1981).

46. Harvey Brooks and Eugene B. Skolnikoff, "Science, Technology, and International Relations," in Philip W. Hemily and M.N. Özdaş, eds., *Science and Future Choice,* Volume 2 (Oxford: Clarendon Press, 1979).

47. William H. Kincade, "Over the Technological Horizon," *Daedalus* (Winter 1981): 105–127.

48. *Science Indicators 1978,* Report of the National Science Board (Washington, DC: National Science Foundation, 1979), pp. 24–26.

49. National Academy of Sciences, Commission on Human Resources. *The Invisible University.* (Washington, DC: National Academy of Sciences, 1969).

50. *Science Indicators 1976,* Report of the National Science Board (Washington, DC: National Science Foundation, 1977), pp. 20–33; also Gellman Research Associates, Inc., *Indicators of International Trends in Technological Innovation* (1976).

51. Brooks and Skolnikoff, *op. cit.,* p. 263.

52. Raymond Vernon, "On Science Indicators 1978," to be published in Volume 1, *Indicators of International Technology and Trade Flows* (Washington, DC: National Science Foundation, in press).

53. Science Indicators 1978, *op. cit.,* pp. 14–16.

54. Executive Office of the President, Office of Management and Budget, *Fiscal Year 1982 Budget Revisions* (10 March 1981): 43–46; also Energy and Science Division, Office of Management and Budget, *Revised Special Analysis of the R & D Component of the 1981 and 1982 Budgets of the U.S. Government* (March 1981): 4–5.

55. Frederick August von Hayek, "The Pretense of Knowledge," Nobel Memorial Lecture, Stockholm, 11 December 1974.

Needs, Leads, and Indicators

Robert S. Morison

What are science indicators supposed to indicate? The obvious analogue, of course, is provided by economic indicators—e.g., growth rate of the Gross National Product, cost of living, rate of inflation, or the percentage of unemployed. In this more familiar case, the numbers are supposed to indicate the state of an entity called the "health of the economy"—but there is a circularity here. The health of the economy is not only measured by these indicators; it is largely defined by them. Indeed, these "indicators" often assume an importance at least as great as that of the abstraction they are supposed to quantify. Over the years, for example, economists, businessmen, and readers of *The Wall Street Journal* have come to feel in their bones that a growth rate of 5% is good, an unemployment rate of 5% tolerable, and a double-digit inflation rate dangerous. Such indicators are regarded as "good" or "bad" in themselves and as things that may be treated directly, just as a physician treats painful symptoms when the underlying disorder that they signal lies beyond his reach.

The analogy between economic and science indicators is obvious but misleading. True, a number of elements in the scientific enterprise can be measured. There are "inputs": trained people, laboratory space, equipment, and, of course, money. There are also "outputs": papers and books, brilliant generalizations, and Nobel prizes. If applied science is included, outputs also extend to material matters such as weapons systems, new drugs, and improvements in man/hour productivity. But how are these variables related to the health of science? Indeed, in the case of science, it is even less clear that an entity with meaning independent of its measuring sticks really exists. Furthermore, we have no background of experience to give us that feeling in our bones about what the indicators say about themselves. No one has any real ideas about what proportion of the scientific population *ought* to be doing radio astronomy, how many papers *ought* to be published each year, or *ought* to be cited by how many Europeans. For these and perhaps other reasons, most commentators on the use and development of science indicators seem unhappy with the rate of current progress. It therefore may be worthwhile to return to square one and ask why we want such indicators at all. May there not be a better way of reaching the presumed objective?

The current interest in science indicators stems directly from the somewhat belated efforts of the National Science Board to fulfill the Congressional mandate to investigate and report periodically on the state of the scientific enterprise. The Congress was not prompted by idle curiosity. It clearly wanted help in deploying and financing the nation's scientific resources.

Science policy—or, more crudely, the allocation of money for science—can be divided into two quite different activities: (a) *the strategic deployment of funds* among various areas or categories, and (b) *the tactical distribution of appropriated funds* to particular programs or projects within strategically selected areas. Much thought and administrative effort have gone into arrangements for tactical distribution; almost all Federal government agencies have settled for some variant of the peer review system. Although questions have been raised periodically (most often perhaps about the representativeness, objectivity, and disinterestedness of the peers), repeated evaluations both by committees of "wise men" and by teams of social scientists have given the procedure a surprisingly clean bill of health. Certainly no one, however critical, has come up with a better way. The rather general approval rests on the consensus that tactical allocations should be made on the basis of scientific merit and that reasonable agreement on merit can be reached by appealing to criteria internal to science.

The situation regarding strategic deployment of resources is quite different. Here there is disagreement on the factors to be considered, no good way of measuring them in any case, and not even a clear description of the administrative mechanisms for reaching decisions. We shall not venture any further into the maze than is necessary to identify a few threads of importance from the standpoint of indicators. In the first place, almost everyone except the purest of pure scientists agrees that factors other than scientific merit must also be considered in making strategic allocations. High on the list are the felt needs of the nation for better health, better weapons, more abundant supplies of energy, and so on. Then there are somewhat different needs felt by particular regions to upgrade their scientific capabilities, or by women and minorities for special attention in order to share in the strategic pie.

Because of the recent clamor for public participation in decision-making at this level, one might point out that it is not a new idea. The public has been explicitly represented on the advisory councils of the National Institutes of Health (NIH) since their inception, and the National Science Board usually contains several members appointed for reasons other than scientific expertise. Indeed, it might be argued plausibly that the size and shape of the nation's health research budget is due more to the vision and energy of a non-scientist, Mary Lasker, who has served on many NIH councils and has had close relations with Congress, than to any three members of the scientific community. Finally, of course, non-scientists in the U.S. Office of Management and Budget work over the research budgets for all Federal agencies, which are then voted on by the public's representatives in Congress.

Where in this echelon of decision-making would science indicators be most useful? By and large, they are not needed by peer reviewers concerned with scientific merit. It is the strategic planners engaged in adjusting the scientific enterprise to societal needs who require additional help. And, as soon as we say this, it becomes clear that indicators should tell us something about how science is doing in relation to what it is supposed to do for human welfare.[1] The strategic planners and especially the public's representatives in the Office of Management and Budget and the Congress are interested primarily and properly in goal-directed research. Some goals may be a long way off, and much basic, vaguely targeted research may be needed along the way, but the public eye is on the result. Public money will not be spent without some goal in mind.

The scientific community has different concerns. Although many, possibly most, scientists may be silently pleased to know their work has benefit to society, their explicit worry is that emphasis on societal needs will encourage neglect of basic science. The scientific community, therefore, is always searching for new ways to define basic research, and one strongly suspects that the indicators movement had its origin in the desire to erect more quantitative and objective defenses. So far this hope has not been realized. And I am beginning to doubt that indicators will ever be directly useful in budgetary planning for basic research. As I shall discuss in a moment, it may be easier to indicate where we are in certain areas of applied research. The state of applied research may then be used to give some hints about the need for basic investigations.

The current literature suggests that the difficulty lies not in identifying or even quantifying a number of different variables of the scientific enterprise which could in theory serve as "indicators"; rather, it comes in devising a scale against which to read them. Determining the ranges over which we should feel pleased, displeased, or indifferent is critically important. In other words, what is par for the course? Attention to counting and data processing cannot conceal the basically normative intent of the enterprise, nor the concurrent difficulty of establishing norms.

Indeed, there are many things to measure. In several important cases, the inputs—of dollars, of working scientists, males, females, and Hispanics, graduate students, post-docs, and professors—are well-known and further broken down into industry, government, education, or lost to view. We also have counts of outputs—of papers, citations of papers, and Nobel prizes arranged according to national origin. But how do we know what the numbers "ought" to be? For the most part, we fall back on comparing the present with the past or the United States with other countries, notably the Soviet Union. Presumably we would be disturbed if the number of published papers revealed a steady decline over time or if the Soviet Union began to surpass us in Nobel prizes. On the other hand, such indicators help very little in determining the health of science in any absolute sense or, more practically, in relation to what it might be if organized and financed at some theoretic optimal level.

Not long ago, after a long morning of discussion of such matters at the National Science Foundation, I was jolted by Hugh Loweth into a different way of looking at norms. Loweth has long had the principal responsibility for science in the Office of Management and Budget. He is well-informed, sympathetic to science in general (although critical of some of its habits), and ironically aware of the constant mismatch between science and politics. In this instance he waited silently through a very long morning to finish things off with a Parthian anecdote.

It seems that when the science budget was first presented to President Carter, he began his review by asking his Science Advisor if anyone had ever drawn up a list of 200 significant problems facing the United States and then gone on to show how R&D might help solve them. Apparently the Advisor pleaded newness on the job and referred the question to the more experienced Hugh Loweth. What Loweth said when the buck passed to him, he did not relate. He merely flashed an enigmatic smile—such as may have lit the face of Pontius Pilate when he put his famous question about Truth "and did not stay for an answer."

In spite of the ambiguities, my own view is that Hugh Loweth is a serious man. I have therefore brooded over his words and even more over the enigmatic smile. Could one, I asked myself, really discard the habits of a lifetime and *begin an appraisal of science with social needs rather than scientific leads*? Once identified, the needs could suggest goals or targets for science. The next step would estimate progress towards the goals and the research needed to reach them—in other words, to assess the state of science not on some absolute scale but in relation to what society hopes or expects of it.

I shall return to a more detailed proposal in a moment, but first let me say that the notion just outlined may not be quite so bizarre as it sounds. In point of fact, considerable industrial research is already planned and appraised in just this way, as Edward David recently explained in an editorial in *Science* magazine.[2] He began somewhat sardonically by quoting David Sarnoff's statement that 50% of inventing is knowing what to invent, and went on to say that industrial research "must be part of a larger whole, which incorporates development, often involving highly organized project work; engineering design, including means for manufacturing and quality control; securing financial support; and marketing the result in accord with consumer preferences." I should note that David's point was different from the one I am struggling to make. He was primarily concerned with showing that it would be unwise for the government to enter the industrial sector in any serious way because, in the nature of things, it could not carry on the type of closely integrated effort just described. The industrial model is indeed too specialized and perhaps too tied to market considerations to serve as a model for conducting the nation's overall scientific effort; nevertheless, the industrial experience does show that a scientific effort can be appraised in terms of its relation to a defined goal; it seems not impossible that, *mutatis mutandis*, the same approach might be used on a wider stage.

Where does this concept of research as a way of reaching a set of socially determined goals leave us in regard to the appraisal and support of basic research? Perhaps we can begin by admitting that there is no clearly defensible way of deciding purely on its own merits or state of health what a budget for basic research ought to be. This is as true for a single corporation as it is for the nation as a whole. Directors of industrial laboratories have long known this and are quite frank in explaining how they handle the matter. First, the director arranges and defends the R&D budget on the basis of the results likely to be of immediate benefit to the company; then he adds a certain amount for activities that may be reasonably expected to pay off in five or ten years. Then, as a kind of afterthought, he adds x percent for research whose early results cannot be foreseen, but for which experience tells him may be ultimately related to the company's welfare. If he has done it correctly and the overall record of his lab is a profitable one, the budget committee (usually made up largely of non-scientists) is likely to go along. Many of the members rely on little more than the gut feeling that a good director ought to be allowed his five or ten percent of whimsies.

The defense of the national budget for basic research has been carried on in a more explicit, elaborate, and possibly more dignified way, especially since Vannevar Bush's publication of *Science, the Endless Frontier* in the mid-1940s. But, in spite of the rhetoric and the more elaborate rituals, the process is much the same. Congress votes large amounts for targeted research in the military, space, health, and other agencies, and then provides what seems a reasonable fraction to the National Science Foundation for more basic investigations. It has also been willing to shut its eyes if some (but not all) of the mission-oriented agencies support research whose mission orientation cannot be immediately proven. Nevertheless, willingness thus to indulge the scientific community depends in the long run on how the targeted research program is going. When the National Institutes of Health seem to be faltering in the war against specific diseases, then less money is available for the Institute of General Medical Sciences. When man/hour productivity declines, the National Science Foundation is pushed in the direction of applied research. Thus, it appears that the "health" of basic research depends in turn on the perceived success of applied research. It follows that what we really need is a better way of evaluating and managing targeted research.

To orient our search for indicators around applied rather than basic research has at least two additional advantages: (a) it is much more likely to be understood and appreciated by the public and (b) it makes use of the fact that a given bit of applied research has, by definition, a clearly defined goal. It is this approach that I now wish to explore. The basic idea is a simple one. *Select a particular goal, make an estimate of how far we are toward it, and assign a numerical value to the estimated distance. Record a judgment of the adequacy of existing methods and conceptual frameworks. If inadequacies are found, estimate what is necessary to repair them and develop the ratio of what is being expended to what might be wisely spent. Assign numerical values as may be desirable.*

Some readers will immediately object that the proposed procedure is unscientific because it involves subjective judgments at so many crucial points. The counterargument is that objectivity and precision often must be sacrificed in order to gain significance. Since it is possible to argue endlessly and pointlessly about this dichotomy, let us turn at once to three real examples of targeted research and see how the proposed evaluative procedure might have worked if it had been available. All are from the same field—virus disease.* Each, although specifically targeted on the control of a specific disease, led to and/or depended on concurrent advances in understanding of sufficient generality to deserve classification as basic research. Two have reached their specified goals; the third may be regarded as still unfolding.

Yellow Fever

In 1951 when the Rockefeller Foundation decided to "give aid in the eradication of yellow fever in those areas where the infection is endemic and where conditions would seem to invite cooperation for its control," the project was regarded more as a practical exercise in public health than as a research program; but the emphasis soon shifted towards research. It was known (or thought) that yellow fever occurred largely in urban areas at irregular intervals and with varying mortality, often very high. At the turn of the century, Walter Reed had convinced most people that Carlos Finlay had been right in

---

* To avoid misunderstanding I should point out that I am not a virologist, nor did I work on any of these projects. I have been an interested onlooker of all three efforts and have known quite well some of the principal actors in the first two. In this informal account, I rely largely on memory. Experts may find that my emphases and even some of the facts may be wrong, but this should not affect the methodological conclusions.—RSM

surmising some 20 years before that the disease was carried by a mosquito (*Aedes aegypti*). Reed had also shown, apparently quite clearly, that the infectious agent was a filterable virus. General Gorgas and others had shown that it was possible to control the disease in the cities of the Western Hemisphere by controlling the breeding places of *Aedes*.

Eventually it was discovered that the ease of control was due in part to the fact that the *Aedes* strain found in the Southern Hemisphere was a domesticated clone, adapted to man-made rain barrels and to pottery found on ships that plied their disastrous trade across the middle passage. As such, it had not yet inserted itself deeply into the ecology of South America. However that may be, in the early 1900s the job of public health administration to eradicate the disease was thought to be straightforward. If we had then estimated the degree of available knowledge against the apparent need we might have optimistically proposed a 9 on a scale of 10. Almost immediately, however, things were thrown into confusion by Hideyo Noguchi's announcement, backed by the prestige of his mentor Simon Flexner, that yellow fever as seen in South America was not caused by a virus but by a leptospira. In addition to confusing the local situation, it raised a further question: was the disease known as yellow fever in this hemisphere the same as the one encountered in Africa under the same name? The "adequacy of knowledge" indicator might well have fallen to 4 or less. The scientific community did, in fact, regard current knowledge as inadequate, for it was shortly decided to send a team to Africa to do a thorough combined field and laboratory investigation to repair the deficit.

By 1927, the agent had been shown unequivocally to be a filterable virus, transmissible to monkeys. The indicators for nearness to goal and adequacy of knowledge might both have moved then to 6 or 7, but no more than that because researchers still needed some species more numerous and less expensive than the monkey in order to make accurate diagnoses, trace the virus through human populations, identify additional vectors, and (ultimately) develop a vaccine.

By 1930 a young South African research fellow at Harvard, Max Theiler, had shown that the virus would grow in the brains of living mice. This discovery enabled a Rockefeller Foundation team to develop a test not only for the virus itself but for the development of antibodies in exposed persons. The latter was essential for identifying subclinical infections and tracing past epidemics. In the course of these studies, Theiler was asked to join the Rockefeller laboratory, where he embarked on a series of further explorations using animal and tissue culture methods which in other circumstances might well have been characterized as "basic research." At this point, the index might have been raised to an 8 or 9, but just as things were going so well in the laboratory, field studies raised further difficulties. Yellow fever was clearly not confined to domestic environments in Africa where the wild monkeys acted as a permanent reservoir. Now it was shown that the virus had migrated to the jungle in South America as well. Indigenous species of monkeys and mosquitoes were involved and their way of life made it impossible to think in terms of species eradication on a large scale.

The discovery of jungle yellow fever made the development of a vaccine imperative. Sanitary measures alone would not suffice to protect the rural populations. Fortunately, by 1937, the basic laboratory studies which combined unrelenting industry, careful record-keeping, inspired intuition, and good luck resulted in a strain of virus that produced satisfactory immunity in animals and man without doing measureable damage to either the nervous system or the viscera. The procedure used was the serial cultivation of a West African virus strain in a number of different culture media, most of which involved either mouse or chick embryo tissue. The good luck came when one of the chick tissue cultures produced an attenuated strain, apparently by a mutation, which never could be induced again. As a result, all the vaccine used today descends from the single mutant known as 17D. The results of this combination of good luck and hard work would have justified ratings of 9 or 10. All that remained was development of methods for making the vaccine in quantity and testing it under field conditions.

This condensed history tells us several things about evaluating and financing a specific area of targeted research.

1. It is possible to select a target and make more or less continuous estimates of our distance from it. These estimates will never be completely accurate because of the role of chance in discovery. Unexpected turns of fortune can either slow or accelerate progress, but however imprecise, the effort to see both where we are now and where we might be can help in making administrative decisions, as in the Rockefeller decision to establish the New York laboratory for Dr. Theiler.

2. It is possible to identify areas in which something suspiciously like "basic research" is needed, to find the people capable of doing it, and to train more like them. Nevertheless, in the course of basic work something of immediate practical importance may also be found.

3. Administrators and the interested public can talk about targets without interfering with workers who prefer to think of themselves as doing the basic research.

4. Patience and persistence are equally necessary in the support of both applied and basic research. It may take decades to reach the goal even when it is clearly seen.

Poliomyelitis

When Basil O'Connor decided to conquer polio, he was not a popular figure in scientific circles, nor did he ever really become one. His blatantly "applied" approach irritated defenders of basic research, and his dictatorial encroachments on scientific prerogatives often offended professional sensibilities. Even today one could argue that the Salk vaccine was launched into use a year or two earlier than it might (or should?) have been, because of O'Connor's impatience. This mismatch of personal style may have kept the scientific community from learning as much as it might have from this experience in evaluating and planning research.

The entire story is too long to review here,[3] but the following highlights stand out as relevant to my essay topic. O'Connor began by assembling a team of first-rate advisors in epidemiology, virology, immunology, neurology, and clinical medicine. It was an astonishingly well-selected team for what is now called interdisciplinary research. At the outset there was little to go on. Researchers generally agreed that a virus was responsible for polio, but no virus had been isolated and identified; they could only guess at the method of transmission from one person to another. To make matters more difficult, it was not known how the virus reached the nervous system from its point of entry into the body. Did it go through the blood stream as many infections do, or did it travel up the nerves? The mode of transmission was important in part because it was believed that a vaccine would be more successful against a blood-borne than a nerve-borne disease. With all these crucial unknowns in mind, it would have been difficult to justify an index greater than 2 or 3. Because of this lack of basic information, many people doubted the wisdom of a targeted attack.

In retrospect, it is clear that the basically infectious nature of the disease was known, the next questions were clearly defined, methods were available for answering them, and in each area, what proved to be unusually capable investigators were waiting to be involved. Many of these individuals would have regarded themselves as basic scientists, loftily detached from applied research. Some, like David Bodian in anatomy, or John Enders in virology, would have been defined as basic researchers by almost everyone. No matter. In an amazingly short time, work moved forward in both knowledge of viruses in general and practical methods of controlling polio. We might assign a 6 or 7 to the availability of methods and concepts at that point.

It was soon shown, for example, that the virus was carried by the blood and not in the nerves. The epidemiologists found that, in the past, most people were exposed at an early age and became immune for the rest of their lives. The increasing numbers of adult cases in advanced countries in the twentieth century was explained as the result of improved sanitation that had the effect of postponing infection. Both types of observation greatly enhanced the possibility that a vaccine could control the disease if the virus could be grown in suitable quantities. By this time, the goal indicator had surely risen to 5 or 6; and, even if the kind of thinking involved was more implicit than explicit, it resulted in increased budgeting for some of the best basic virologists in the world. In what now seems a very short time (but for those of us with children growing up in the late 1940s, an uncomfortably long one), John Enders and his colleagues had shown how to grow the virus in cultures of human tissue. The indicator could have risen to 8 or 9 at this point because all that remained was to produce a killed virulent virus and a set of living, attenuated strains, and to do the practical field testing necessary to decide which to use where. These last steps were, of course, difficult and time-consuming but, once the knowledge indicator reached 9, inevitable.

Hepatitis

The history of research on hepatitis is much more complex. Although work began on the disease during World War II, we are still far from the goal. The name is now known to cover at least three distinct infections; there are still no very convenient animal hosts; and tissue culture has proven to be unusually difficult to establish. The immunology is so unusual as to be almost bizarre, and the most useful antigen for identification was discovered by a scientist who was looking for something quite different. It would be hard to give a "nearness-to-the-goal indicator" a value of more than 5, but the very exercise of evaluation could help to tell administrators where they ought to allocate funds. A low index, for example, suggests the need for broad distribution of funds at fairly basic levels. In the case of hepatitis, it may be possible to develop an entirely new type of vaccine by utilizing what appear to be fragments of the viral coating, shed into the blood stream of victims of the disease. This finding suggests the need for further basic work on the molecular structure of this coating and how it might be reproduced.[4]

What would be particularly interesting in complex, unfinished cases like hepatitis would be experts' judgments on the subsidiary indicators that describe the availability of ideas and methods, and the budgetary backing sufficient for current research on the subject. Thus, estimates giving a nearness to the goal of 5, an availability of ideas and methods of 9, and a budgetary index of 4 would suggest not only where we are, but also where we might be and why a 5/5/10 distribution would hold a different lesson for the budget officer. Some critics might complain that the national cancer research program is in the latter category.

Conclusions and Extensions

How can these relatively simple principles, derived from simple individual cases, be extended to the nation's research program as a whole? Much of the planning within single agencies is already done implicitly in this way. The National Institutes of Health, for example, have been particularly successful in assessing the state of knowledge and the availability of personnel in given areas and then mounting programs to repair the discovered deficiencies. The organization has not been so good at assessing the state of health research as a whole. Research effort is unevenly distributed. There are very large expenditures for cancer research of certain kinds yet only a surprisingly small program to study the effect of environmental contamination on health, including possible carcinogens. In fact, the entire environmental health effort, except in the classical areas of infectious disease, probably deserves a score of 4 or less. Although its methods of study need refinement, much research can be carried out with the means already at hand. Environmental health research might get a three-part score of 4/6/3, indicating that we don't know as much as we could or should and that the fault lies with those responsible for planning and financing, not primarily with the state of the art, although this does need substantial improvement. There are other more subtle difficulties. Much environmental research is boring, unfashionable, and time-consuming; it does not have the same self-starting dynamics of fields like molecular biology or particle physics—but that perhaps is another story.

By now it should be clear that, whatever the merits of the indicators I have been discussing, they tell us little about the state (or "health") of science as a whole. Indeed, one begins to suspect that there is no such thing. Perhaps the best we can do is to select a series of goals and ask ourselves how close we are to each one. Those so inclined could then sum these to produce a single global figure, although there is no reason to suppose that such a "whole" would be any greater than the sum of its parts. Nevertheless, those responsible for deploying research resources where they can do the most good could be guided by the pattern of the individual indicators, as suggested in the comparison of cancer and environmental research above.

All my examples so far have been drawn from the health area because these are the fields in which I am least ignorant. However, the same approach may be useful in other areas. For example, there are many reasons for believing that we will burn increasing amounts of coal during the next century or so. To burn it safely we need to know more about:

1. Where and how much coal there is;

2. What potential contaminants the different deposits contain, not only the notorious sulfur with its acid rain but also heavy metals like mercury which are difficult to remove and toxic in very small amounts; and

3. The "greenhouse" problem. Current ignorance involves not only uncertainty about the rate of increase of CO_2, but perhaps more unfortunately, also the current and potential role of tropical forests and photosynthetic processes in the ocean.

These are obvious needs of the greatest social importance. Methods ranging from routine surveys to complex ecological analysis are available for studying most of these matters, but are not being used on anything like the necessary scale. We might give the research effort on the effects of burning coal a three-part score of 2/6/1.

Social science research has also identified many inadequately studied problems. "Why Johnny can't read" is one that, for example, would require a massive research effort as well as possibly a reorientation in educational philosophy. Everyone knows how important the problem is. Almost everyone has opinions about it. Every year, large sums are appropriated for remedial procedures, but (so far as I can determine) controlled studies of alternate methods of teaching reading are simply unavailable on the appropriate scale. The last ten years have seen enormous progress in the design of large-scale comparisons of different therapeutic regimes for the treatment of such abnormalities as hypertension, myocardial infarction, various forms of cancer, and most recently, certain psychotic disorders. Many of these studies extend over periods of time considerably longer than it takes to teach people to read, and the expected outcomes are often no easier to measure than reading comprehension. With these valuable examples in mind we might give the following index numbers to the state of research on reading: distance toward goal 2, availability of methods 8, organization of effort 1.

Another important case for a society based on the concept of equal opportunity is that described by the titles of Christopher Jencks' books *Inequality* and *Who Gets Ahead?* [see the review in 208 *Science* (16 May 1980): 707]. Here is a problem of overwhelming social importance about which very little is known, for which suitable methods are at least becoming available, and for which there is at least a small cadre of competent investigators. But the overall effort is small relative to other areas of much less social consequence. Again, a score of 2/7/1 (or thereabouts) seems in order.

In summary, I found the "Loweth proposal" somewhat more fertile than I expected. It certainly isn't the only way to evaluate science—and it may not be the best—but it does call attention to areas significant to the general public, areas that are scarcely touched by the "indicators" so far proposed. Neither the number of Ph.D.s granted each year, nor the count of American papers cited in Swedish journals, directs our attention to the fact that there is only one research station in the world that can tell us anything useful about the long-term trend of carbon dioxide concentrations in the air. A long string of Nobel prizes in molecular biology serves only to distract us from such uncomfortable realities as the lack of solid evidence on which to base the atmospheric standards for nitrogen oxides that cost American automobile owners billions of dollars each year. Perhaps the most important lesson is that the "health" or "welfare" or "competence" of science cannot usefully be appraised as a global entity for which appropriate global indicators can be found. Science is very much the sum of its parts and if some of these parts fail in development, then the whole is proportionately diminished. Our proposal deals primarily with those parts of science which seem to lend themselves most easily to numerical indicators. It would, of course, be a mistake to evaluate the entire scientific enterprise purely on the basis of how it answers the question "What have you done for us lately?" Especially at the basic level, provision must be made for the continued inner direction of science. It is the self-propelled excursions into the unknown that have uncovered entirely new ways of ordering the world around us, reduced superstitition, and in general improved the quality of life. The fact that the unforeseeable nature of these advances makes it difficult to devise "indicators" and rating scales should not be allowed to reduce our concern for them as we turn attention to rating our progress towards more definable goals.

Notes

1. When this paper was presented to the Harvard-MIT faculty seminar on indicators of quality in science (1979–80), several discussants showed an interest in hearing more about possible mechanisms for determining the social utility of particular projects or programs. This subject lies beyond the scope of a paper on science indicators *per se*. For our present purposes, the significance of social goals lies only in their role as endpoints to a proposed scale of measurement.

2. David, E.E., Jr., "General Sarnoff and Generic Research," editorial, 207 *Science* (15 February 1980): 719.

3. Saul Benison describes the progress of this research in his essay "The History of Polio Research in the United States: Appraisals and Lessons," pp. 308–343 in Gerald Holton, ed., *The Twentieth-Century Sciences: Studies in the Biography of Ideas* (New York: W.W. Norton and Company, Inc., 1972).

4. Since this was written, rapid progress has been made not only in isolating the molecular fragments from human blood, but in characterizing and cloning the DNA responsible for the synthesis of the viral envelope. See 213 *Science* (24 July 1981): 406.

The Quality of 'The Quality of Science': An Evaluation

Bruce Mazlish

I

The quality of science? The inquiry seems paradoxical, for science prides itself on being quantitative. The 25 September 1979 U.S. Comptroller General's Report on Science Indicators cites many examples supporting this fundamental position, from Kepler's remark that "... the mind comprehends a thing the more correctly the closer the thing approaches toward pure quantity as its origin" to Lord Kelvin's "well-known belief that if something cannot be quantitatively measured it cannot be understood."[1]

In assessing the quality of science, the inclination of the natural sciences is to establish quantitative measures—that is, science "indicators." Surely, if Bentham and the Utilitarians believed that they could measure something as vague and ephemeral as happiness by setting up a "felicific calculus," how much more humble and realistic should appear the hope of measuring the "health" or "quality" or "happiness" of science by establishing numerical markers? Science's own scientific nature, therefore, supports the effort to understand the quality of science in a Kelvin-like fashion. The paradox seems resolved by definition.

The value of science indicators appears self-evident.* The limits and limitations involved in the approach are less obvious. Although I believe firmly in the help afforded by science indicators, correctly used, I believe that the help is limited, can carry with it unintended

---

* Incidentally, most of the time when I write "science," I am discussing both science and technology. As with the term "science indicators," I am simply employing shorthand. Occasionally, for emphasis and as a reminder, I write "science and technology." A future task of work in this field should be to introduce meaningful discrimination between the two terms.

harmful possibilities, and may cause us to overlook or undervalue other means of understanding and even "measuring" the quality of science.

The implied paradox of quality and quantity is only one of our problems. Another is the conception of what is to be measured. What is the object of our study? The "health" metaphor that came into early use may have been an unconscious carry-over from efforts to measure and support public health R&D, or may be inspired by notions about a community's "health," about the "body politic." It seems natural to discuss "taking the pulse" of the public through opinion polls, etc.; therefore, why not do the same with the "body scientific"? Such analogic thinking is basic, and can be very valuable; but, as I have remarked elsewhere, there are also dangers:

Historical analogy gives flesh to perception of vague resemblance. It is not a rigorous form of reasoning, but it is one of the more attractive. It is, too, a fashioner of myths—durable ones that survive, like a locust's brittle armour, even after life itself has departed. Analogy, finally, has but one eye, and it sees only similarities.[2]

The body analogy and the health metaphor, although appealing, have been found useful only up to a point. In 1968, Congressional legislation, P.L. 90-407, mandated an annual report on "the status and health of science"; and Donald F. Hornig, writing in *The Health of the Scientific and Technical Enterprise*, a report to the Office of Technology Assessment, carried the approach as far as it will profitably go, consciously invoking the "analogy to medical diagnosis."[3] The Comptroller General's Report bluntly stated its objection to this approach:

...much of the terminology generally used to develop measures (e.g., 'health' and 'vitality' of science) is vague and evaluative. (p. i)

and

Good 'health' is a metaphorical standard, one which must be *interpreted*. While the health of a human being is generally well-defined and understood, for science and technology there is no standard or accepted definition of a healthy state, or even agreement if such a state exists. (p. 17)

The Report concluded that "It would be at least of some benefit if nonevaluative terms such as the 'state' or 'condition and direction' (of science and technology) were used."[4]

"State" of science and technology *seems* a nonevaluative and unmetaphysical term that ought to lend itself to neutral, objective, statistical measurement. Closer examination, however, shows the matter not to be so simple. As the Latin root indicates, "state" is from *status* (which also means condition) and, thus, relative standing. Standing in relation to what? The preceding condition? This is to imply an infinite regress.*

Somewhere in this mélange of terms and meanings, it becomes clear that while one can minimize the metaphysical and analogical, the valuative element in the endeavor remains. This is not to say that the effort to construct and to use putatively value-free science indicators has not been, and is not being, made; quite the contrary. It *is* to say that such an effort—in my view, self-deluding—is in any case now directed most frequently at establishing measures for the *quality* of science. And the resort to the term "quality" is clearly evaluative, although in what sense remains to be seen.

This leads to our third problem. If we accept that the object of our studies is the quality of science, how are we to understand that term? Here, it seems to me, there is an inside and an outside usage. Professional scientists—i.e., those engaged in the actual practice of natural science—regard quality assessment as some professional estimate of the worth of a particular piece of science, as in the statements "Physics at X University is done well" or "Professor Y is noted for his elegant and excellent work in spectroscopy" or "Dr. Z is engaged on an inconsequential topic and his findings are highly suspect." Scientists on the "inside," therefore, believe they can establish quantitative indicators that reliably measure the quality of science (or its state, or even its health). We shall return to the "inside" view later.

Another meaning of quality, however, concerns itself with the "outside," that is, with the impact of science and technology on society. The concept *quality of science* in this sense is obviously connected to the concept *quality of life*. Earl Stevenson, Chairman of the American Academy of Arts and Sciences' Committee on Space, in his Foreword to Raymond Bauer's *Social Indicators*, makes this explicit when he remarks that "Such measures of social performance are all

---

* As for *statistics*, that means quantitative data, with its Latin root, *statisticus*, reminding us that it also means statesmanlike (see, for example, Funk and Wagnall's *Standard Dictionary*). Both roots point to the political nature of the term; and it is no accident that state means "body politic" as well as "condition" or "status."

the more important in a 'post-industrial' society, one in which the satisfaction of human interests and values has at least as high a priority as the pursuit of economic goals."[5] Stevenson might have added military goals as well, for there the issue can be put dramatically: Scientific research and development of poison gas or bacterial warfare might be of the highest "quality" judged from the "inside" and the lowest "quality" judged from the "outside" (i.e., in terms of society's human interests and values).

Both inside and outside measures of quality are needed, with the former subordinate to the latter. There can be no substitute for evaluating science in terms of its impact on society—society and technology are, after all, *human* endeavors, they are great testaments of human creativity—and thus, for looking at science openly in evaluative terms. Most scientists (any more than most generals) are probably not happy with the conclusion that "Science, like war, is too important to be left to the scientists." Fortunately, however, one can count on the fact that, for many (the General Marshalls of science), dedication to truth will prevail in the end.

At this point, let me say a brief word about the *nature* of the natural sciences, social sciences, and humanities. In a provocative article, "Scientific Concepts and Cultural Change," Harvey Brooks asserts that "Art, scholarship and science are united in looking further behind the face of commonsense reality, in finding subtleties and nuances."[6] He is surely right about their looking behind so-called commonsense reality and, at one level, about their sharing a common desire for subtlety and nuance. At another level, however, science and humanities are propelled by different impulses and needs. Science seeks to reduce matters to their simplest terms, to group such simplified phenomena under a single law, and to rid itself of all ambiguities. Its favorite mode, correspondingly, is the quantitative and precise; and its aim, generalized law. The humanities, on the other hand, wallow in ambiguities and ambivalence and, if anything, seek to pile additional meanings on an outwardly simple datum. Complexity is not only the bedevilment of the humanities, but also their hearts' secret desire. Rather than try to reduce the number of variables, humanists constantly try to increase them, sensing the interconnectedness of all things and aware of the fallacy of misplaced concreteness and the hubris of partial knowledge. The social sciences, in turn, are attracted both to the generalizing and simplifying aims of the natural sciences and to the value-oriented and subjective interests of the humanities.

This is an immense and important topic, impossible to approach in depth here. What has been said, however, ought to indicate the spectrum on which questions about the quality of science are likely to be sighted. From the outside—i.e., in terms of its relation to quality of life—the quality of science cannot be seen free from value, a value which is necessarily connected to human values. It follows that the natural sciences and the humanities, with the social sciences, must all join their different perspectives and methods in attempting to deal with the subject.

II

The effort to quantify the quality of science has become a specialty in its own right, a professional sub-field of science (or of the sociology of science, dependent on one's view). Scientometrics has its own international research journal, and practitioners number anywhere from 200 to 400. In Derek de Solla Price's optimistic words,

> Progress in the provision and analysis of science indicators has now reached the point where we have accurate, reliable and reproducible data for the output of research in science and technology by quantity and quality [i.e., what I have called "inside" quality].[7]

I shall not attempt to say much about the details of Price's own work, such as his construction of science indicators for the major nations of the world and for the major substantive fields of science. His colleagues in scientometrics may argue over details of procedure and computation, but I am prepared to concede that his data yield clever, interesting, and important results.* What I wish to concentrate

---

* The use of published articles as one way of indexing scientific activity takes on added conviction when we place it in the context of Robert K. Merton's reminder that the use of systematic references and citations is itself a new tool of science. Such use subjects scientific assertions to public testing as well as serves the social purposes of securing the scientist's intellectual property by the very act of making it freely available to others. Without direct financial reward, the system offers institutionalized incentives for the simple reason that recognition by qualified peers becomes the basis for other forms of extrinsic reward (salaries, rank, awards, etc.) [Eugene Garfield, *Citation Indexing*, New York: John Wiley & Sons, 1979, pp. vii–viii]. Here we have an excellent example of how sociology supplies

on are the assumptions behind Price's scientometric work. "We need," he tells us, "a social scientific equivalent of the Newtonian masterstroke that took such vaguely used terms as *force*, *work* and *energy*, redefined them with simple equations such as $P = mf$ and $E = 1/2\ mv^2$ and brought order in the previous meanderings."[8] Though Price tries to qualify his statement by prefixing the phrase "Somewhat cautiously it may be suggested...," his statement is very *incautious*. Historians and social scientists are constantly being urged to produce the Keplers and Newtons of their field; but Robert Merton, for example, has pointed out the pitfalls involved in such a manner of thinking.[9] What is important about Price's bit of hubris, hardly important in itself, is that it reveals a mind-set that is scientistic, or pseudo-scientific, in regard to what can be done in the social sciences.

Somewhat more cautious is Price's appeal to econometrics, as a prior example of success in social science. He dismisses the "nihilistic arguments" that claim it foolish "to suppose that the wealth and happiness [shades of Bentham!] of mankind could be computed by counting dollars and pounds."[10] Echoing the claims of econometrics, he urges that his comparable data be used as "a guide for fine tuning." The aim of such tuning is to determine levels of research funding, with the intent of removing gradually "such deviations from the world norm as cannot be justified by reasons extraneous to the cross-national flow of research knowledge."[11]

Price is a gifted and competent scholar. What he has to say about establishing the quality of science from the inside (whatever the necessary caveats) is at the highest level of accomplishment. It is exactly for this reason that his remarks are so important and revealing, and must be taken seriously in the light of concerns for the quality of science, more broadly and externally conceived.

There is no question that a key inspiration for science indicators has been the construction of economic indicators. The Comptroller General's Report cited earlier begins by acknowledging the origins of science indicators in economic thinking: "The Joint Economic Committee requested, as part of their Special Study of Economic Change, that GAO examine the measures used to indicate the state

---

the requisite conceptual setting, which validates the use of science indicators. [For more on such use, see, for example, Derek de Solla Price, "Multiple Authorship" (Letter to the Editor), 212 *Science* 4498 (29 May 1981): 986.]

of U.S. science and technology."[12] Allocation of funds, manpower needs, educational requirements in science and technology—to make policy in these areas, Congress and the President appear to need the counterpart of an *Economic Report* and *Economic Indicators*.

And so, to an extent, they do. But are there limitations and drawbacks in the very model of economic indicators itself that should serve as a caution for those engaged in trying to understand the quality of science, from inside and outside? Recent experience with "fine tuning" the American economy should give anyone pause. A fundamental examination of the very bases of economic assumptions, by such an "insider" as Lester Thurow, suggests what the limitations are to current econometrics.[13] Thurow's challenges (friendly, one might add) are on the highest level of theory. More concretely related to indicators are another scholar's comments on the Consumer Price Index (CPI). He remarks,

Because wage increases, Social Security benefits, and other forms of transfer payments are tied to the CPI—and because economists have tended to foster the myth that the CPI is a totally 'objective,' real number—this measure of change in the price of a fixed basket of market goods is regarded as a true measure of a person's cost of living. More than that, it is frequently extrapolated into a 'rough guide to what is happening to American families overall,' as recently stated by Alan Greenspan, former Chief White House Economic Advisor. The technique of measurement, as well as the underlying logic and reality of the CPI, ill suits it to the claim of being a 'cost of living' index. It systematically overestimates inflation—and by so doing contributes to its escalation; it is based on outmoded data on consumer buying habits; it fails to reflect the differential effects of inflation on different segments of the population; it does not take improvements or debasements of product quality into account; it is full of subjective assumptions—yet one researcher recently argued that it is better to suppress the index's subjectivity than to reveal it and possibly destroy public confidence in the measure.[14]

Such a comment points to the difficulties—and dangers—in designing an indicator, both from the inside and the outside perspective. *Mutatis mutandis*, science indicators must be examined with the same challenges in mind.

Economic and science indicators also are concerned with very different objects: economic indicators such as the GNP and the CPI are not measuring the performance of industries, or groups, with an eye to funding or not funding them, but are measuring the results of "free" market operations in terms of gross total; science indicators, presumably, are measuring the quality of individual, group, or disciplinary activity in an academic, industrial, or governmental funded operation. To put it another way, economic indicators measure the activity of a *market*, to which come agents who already have raised funds in some manner, without asking "should these agents be funded?"; while science indicators assess the world of individuals, even when compounded as a field, say chemistry, in order to decide whether to fund them in the future. Such differences in object (and others) must at least be considered and examined before following the econometric siren too far.

In the end, as in the beginning, concepts and data cannot be separated, nor the social setting forgotten. In a remark several years before Price's "let's not waste time," Bauer commented that

If we examine the President's major policy documents, particularly the Economic Report and the Budget Message, we find practically no information whatsoever on 'social structures.' We find that the major indicators deal not with how good but how much, not with the quality of our lives but rather with the quantity of goods and dollars. This continuation of 'economic Philistinism' is exacerbated by the increasing emphasis upon cost-benefit analysis (often used as a way of releasing resources for Great Society programs), operating on the premise that any meaningful benefits from government programs can be expressed in dollars and cents.[15]

The great temptation, of course, is to turn matters over to the experts, the "insiders." Harvey Brooks once wrote that "Much of the history of social progress in the 20th century can be described in terms of the transfer of wider and wider areas of public policy from politics to expertise"; then, in an extension of this truism, he spoke in praise of "the decline of ideology," or what Daniel Bell and others heralded as the "end of ideology."[16] Brooks was writing in 1965, and I doubt that he would phrase things the same way today. The attempt to restrict affairs to the experts and to "manageable" problems is now seen to be narrow and productive of further unforeseen, often unmanageable problems. (Brooks, incidentally, had the wisdom to caution, even in 1965, that "It remains to be seen to what degree this new reign of the bureaucrat and the expert ... represents a temporary cyclic phenomenon resulting from unprecedented economic growth and the absence of major social crises."[17]) As Raymond Bauer perceived, "An action is often labeled 'technical' to offer reassurances as to its limited implications. The very phrase 'purely technical' is used to convey the notion that an action does *not* have widespread ramifications."[18]

In fact, as we now know, science and technology as well as economics are technical, but not "purely technical," matters. They are also social in nature, with major social ramifications. As "technical" matters, they require careful technical design of their measurements. One must discriminate, for example, among opinions, attitudes, and ideologies in measuring public feelings about science, as Jon D. Miller and Kenneth Prewitt do in their report to the National Science Foundation, about the differences between science and technology, and about short- and long-range factors and results.[19] But as "social" matters, one must always be aware that real understanding comes only in terms of the causal connections, the social structure in back of the data, as a simple example will show. Crime series data over the period 1950–1960 show a rise in certain categories, e.g., in thefts of automobiles and other items worth over $50. The true interpretation of the data is hard to come by, however, unless one also realizes that increasing affluence means that there are more autos and other items valued at over $50 available to steal.[20]

The same connections will be true for science. Data on science must be seen in the actual social (which also means human) setting, and must be interpreted in terms of the full social context in which they are imbedded. Science indicators cannot measure the quality of science in a trustworthy manner unless they themselves are "measured," i.e., evaluated, in turn. Science indicators must be viewed, not as value-free instruments, which exist independent of any larger context, in a vacuum, so to speak, but rather as taking their own meaning and significance from, and contributing to, a larger study, the sociology of science.

III

The indicators movement originated in an effort to study social trends in the 1930s, particularly through a Presidential committee report, *Recent Social Trends,* published in the bleak year 1933. The report had little effect on policymakers, but it did itself establish a trend. Economic reports and indicators that were more limited and policy-oriented followed, with more ascertainable results. As we have seen, economic indicators, pushing aside the more ambitious social studies, became a model of sorts. A return to the earlier inspiration has surfaced in the recent effort to construct social indicators equivalent to the economic ones, an effort spurred by a NASA-sponsored study undertaken by the American Academy of Arts and Sciences in 1966.[21]

The National Aeronautics and Space Administration, in a responsible and highly conscious manner, had wished to detect and anticipate the secondary and even tertiary impact of its efforts to fulfill its primary mission—putting a man and vehicle in space. It quickly became clear to Raymond Bauer, who was in charge of the Academy study, that the inquiry's scope had to be broadened. "Though our interest originated with the problem of measuring the impact of the space program," he wrote, "the problem of measuring the impact of a single program could not be dealt with except in the context of the entire set of social indicators used in our society."[22] Bauer also made it clear that what was involved was not prediction, but anticipation. "One can set goals and make plans, but the cybernetic model demands an active information system with sensors to determine the consequences of actions. In addition, it demands provision for feeding this information back to decision centers and readiness to change one's behavior in response to signals of errors being committed."[23] Thus, human choice and action are not only allowed for, but mandated.

In making choices, Bauer reminds us, one must distinguish two aspects of anticipated events: 1) the probability of their occurrence, and 2) their importance if they do occur. For example, an event may have a relatively low probability of occurring, and yet be cataclysmic in its importance. Estimating and evaluating such matters is often carried out today in terms of the concept of "risk." And the role of human choice in deciding about risk is clearly evident.

Bauer's group did not itself construct social indicators. It did lay the theoretical foundation for their construction, insisting that they should measure *trends* and be situated and studied in a real social context. By 1976, a United Nations statistical office survey "documented no less than 29 countries with social trend books published or in operation." What is more, "a recent UNESCO-sponsored conference on definitions of quality of life reached a remarkable level of agreement between Eastern and Western countries on the necessity of combining evaluations of 'objective' life conditions and 'subjective' values, wishes and aspirations." The ambitions of Bauer and his colleagues seem to have borne good fruit.[24]

One other offshoot of Bauer's committee is suggestive for our discussion of science indicators, as a part of social indicators in general. It is the construction and use of historical analogies as a device of anticipation and as a complement to quantitative measurements. It was my privilege to be enticed into undertaking such an effort. (I had cautioned Bauer that historians, no less than politicians, were always invoking analogies but with little understanding of the methodological problems in such a conjuring act.) In relation to one "piece" of science and scientific impact on society, the space program, I recommended that the most useful analogy to be made was not with the Age of Discovery, but with a comparable "social invention," in this case, the railroad.* (For details of that attempt, see my book, *The Railroad and the Space Program: An Exploration in Historical Analogy*.[25]) It is, of course, simply one other way of trying to anticipate, in this case based on past large-scale experiences. It is also, clearly, relevant only to quality of science from the "outside," that is, the impact of science on society at large.**

Akin to historical analogy in its effort to understand the quality of science by means other than the construction of science indicators is the use of case studies. For example, a case study would actually look at the way funding decisions were made in regard to particular pieces of science or technology—for example, a decision to undertake cancer research, or to build the SST (or not). Such a study would shed light on the quality of science from the inside. By also looking at the subject in a social context larger than the mere funding one—i.e., also studying the eventual impact—the case study would also pertain to quality of science from the outside. Such a study would be best carried out by a team consisting of at least a natural scientist, a political scientist, and an anthropologist; the use of the latter two in constructing science indicators is also recommended.

* I defined a "social invention" as one that is "technological (e.g., missiles, launching pads), economic (e.g., involving largescale employment of manpower, widespread use of materials), political (e.g., involving new forms of legislation, and new dispositions of political forces), sociological (e.g., affecting kinship groups, communities, classes), intellectual (e.g., changing human views of space and time), and so forth.
** For a further discussion of the Age of Discovery, instead of the railroad, as a historical analogy, see my article, "Following the Sun," *The Wilson Quarterly* (Autumn 1980) and my paper, "The Idea of Space Exploration," delivered at the Conference on the History of Space Activity, Yale University, 7 February 1981.

The 'Quality' of Science

A number of different instruments or devices, therefore, may be found useful in attempting to estimate the quality of science in its two senses. Science indicators, constructed with due regard for the dubieties associated with economic indicators, and in line with the best thought about social indicators, are of prime importance. Historical analogies offer both the vaguest and the broadest perspective, gaining in amplitude what they lose in preciseness, alerting us to possible as well as probable connections missed by more certain statistical approaches. Analogies give us primarily a feel for secondary and tertiary impacts on society; methodologically, however, they are very tricky instruments, subject to facile misuse.* Case studies give us the immediate human decisions, the nuts and bolts of specific pieces of science or technology, and are invaluable in reminding us of the nitty-gritty reality behind our clean figures and our soaring analogies. Each approach can supplement and serve as a check on the others.

IV

Science, as any other human activity, is not a static "thing" but a "process." Using science indicators without bearing this fact in mind is a form of mental rigor mortis, emptying the subject of actual life; it becomes the arcane art of the taxidermist instead of the living science of the ethnologist. The Comptroller General's report makes this same point in more prosaic terms:

Quantitative measures alone offer only a partial definition of phenomena. Quantification is a crucial limitation in the realm of social measurement because the complexity of most social interactions and change can rarely be understood purely in terms of quantifiable parameters.[26]

Quite specifically, the report indicts *Science Indicators 1976* for assuming that science and technology "can be measured in terms of resources such as funding and 'stocks' of personnel and a few tangible

---

* The Miller-Prewitt report cited in Footnote 19, for example, is sophisticated; yet, closer examination of its analogy of science to foreign policy, and, consequently, its borrowed assumption of an "attentive public," would be useful: does the sort of accumulated knowledge found in science, as well as the division into basic and applied science, and into scientific and technical, find its counterpart in foreign policy, and if not, does it matter?

results such as innovations and published papers. The process involved in doing science and technology are [sic] not important." And it concludes that, "Basically, this support service model, with its input-output stress, is a technological view of science—one that ignores the internal sociology and conceptual progress of science and technology and emphasizes its economic aspects."[27] The authoritative support of the Comptroller General's report underlies, therefore, the importance of conceptualizing science and technology *as processes*. It follows that, for their fullest understanding, one must employ process-oriented methods and concepts derived from, say, the social sciences and humanities.

One such method, as I have already suggested, is to use historical analogy to enlighten anticipations. Another is to employ case studies. Even within the construction of metrics, however, awareness of process can and should inform our procedures. For example, to measure the public's attitude toward science or a particular piece of science, one can use either a static or process model. The static model asks fixed questions: "Do you approve or disapprove of science?" or "Do you approve or disapprove of the construction of nuclear power plants?" These questions give us precise but limited information. Of more value is what has come to be called "attitudinal polling," which has hitherto been used mainly for measuring public attitudes toward major political events and issues. As Bauer remarks, "The classic before-during-after research in the social sciences has been the panel studies of Presidential elections. In these panel studies, the same individuals are studied over a period of months."[28] In the 1980 Presidential election, the method was extended by pollsters such as David Haskell Sawyer into studies of a panel as a group interacting over a period of time.[29] Daniel Yankelovich's application of this approach to science and technology issues, with sophisticated improvements, is highly promising.* It seeks to combine the best insights

---

* Yankelovich's pilot study (conducted through the Public Agenda Foundation under subcontract to Harvard University) is designed to identify the criteria people use in assessing the impact of science and technology on the quality of life. The research assembles small groups of respondents and presents each group with an array of six specific scientific and technological proposals. Respondents are asked to assess the proposals according to whether or not they feel they should be funded, and then discuss the proposals with the moderator, until all pros and cons have been aired and perspectives fairly presented—a process Yankelovich calls

involving both static quantification and process. It allows us both to witness a model of on-going, creative, social interaction, and to anticipate trends.

A crucial part of process involves values. In deciding what to "do" in science and technology, if not in actually "doing science," attention must be paid to the actors involved and their desires. As Bauer remarks dryly about a related matter, "On what metric do we compare a given level of unemployment among unskilled, young, Negro men with a given level of increase in productivity?"[30] It is easy to work out similar value-laden choices in measuring science and technology. Yankelovich advises looking at public attitudes about what kind of society we should be living in, as a guide to answering the question of what kind of science we should be doing.

The process of choosing is, in fact, a political process. To pretend that decisions about science and technology are purely "technical" is to misunderstand the nature of social reality, of which science is a part. Kenneth Prewitt's discussion, in the Indicators of Quality in Science seminars,* of the "attentive public" for science and of a "scientific elite" assumes that the political process in which they function responds unevenly to the voices of the activists and to the mass of possibly concerned but politically inactive citizens. Yankelovich points out that American science and technology functions in a democratic system in which the public *should* be involved—he favors discussing a widespread "informed public" rather than a narrower "attentive public"—and, in any case, *is* involved.

During the seminars, Willis Shapley commented that the crucial question is actually how people will behave in the face of certain kinds of leadership. The fact is that a scientific elite does exist and does lead. But it is not a unified elite; various members hold different views, and even different values. These leaders compete with one another in determining questions and choices that involve both the

---

"working through." After the discussion, respondents are asked to rank the proposals. Yankelovich's panels, where particular individuals on particular panels may sway its members, simulate the process of leadership in the larger society. His panels, therefore, are microcosms of the larger, public macrocosm, whose leadership is studied by the disciplines of political science and history.

* Kenneth Prewitt, "Current Public Attitudes Toward the Quality of Science: Results from the *SI78* Survey," presentation to the Harvard-MIT faculty seminar on "Indicators of Quality in Science," April 1980.

inside and the outside of the quality of science. They also "compete" (sometimes in co-opting and co-opted fashion) with other leaders: with the President and his Office of Science and Technology Policy (OSTP)*; with the Congress; with industry; and with public interest groups.[31] In short, the process of science and technology, especially today, is fraught with political meaning, and it should be studied in these terms. We can see more clearly what is involved if we recall that, in the early, halcyon days of the "scientific revolution," research was the work of gifted amateurs, operating on a shoestring, frequently out of their own pockets. Academics and those working in quasi-academic laboratories, not those employed directly in the government or industry, were in control of science. Even with increasing professionalization and specialization, knowledge was valued as much for its own sake as for its material benefits.

World War II changed all this—a change that was underway in any case. Huge sums became available for research, the government became the prime source of funding, and applied science began to play a dominant role. For a few decades it seemed as if science policy—and certainly the politics of science—was unneeded, in the sense that money was available for almost all worthwhile projects; it seemed that one didn't need to choose deliberately, and scientists hardly needed to compete with one another for funds.

In the late 1960s, economic realities and the public mood changed. Even putatively good science now had to compete with other good science for scarce resources. As John Ziman reminds us, the long, drawn-out conflict over research and development in nuclear physics "demolished the notion of a united front for science." In the case he cites, the question of British participation in the construction of a new 300 GeV particle accelerator at CERN, "chemists split with physicists on the fundamental issue of the benefits to be derived from the project."[32]

Worse, some of science and its products did not seem so good, and the nuclear power and environmental protest movements came into being. Over a broad front, then, the battle of science was on,

---

* As is admitted, the President tends to be little interested in long-range science policy, but operates mainly on the basis of short-term policy, heavily influenced by political considerations; in the worried words of Leo Marx, the OSTP is "identified with the apex of political power in this country—the White House—and that might not necessarily be such a great thing for science."

with scientists arguing with other scientists inside the "establishment" and science itself having to justify its case with the general public and the politicians and government officials who held the purse strings. The political and social involvements of science could no longer be denied or hidden in shadow.

A few examples will suffice to highlight the situation. In the late 19th century, for example, vigorous discussion took place over the legitimacy of physiological research involving vivisections. Was it legitimate to inflict pain on helpless animals (which might be done in the name of quality "inside" science) because such action helped society at large—i.e., contributed to the relief of man's earthly state by eliminating disease and discomfort? In this case, the answer was "yes." "Inside" science and "outside" science coincided for man, if not for the animals experimented on.[33]

Let us take another kind of case. How, for example, when only limited research resources exist, to determine whether and how much to fund cancer research as against research in auto accident prevention? As Robert Morison makes the calculation, the statistics show that more people die of heart disease and cancer than anything else. In fact, however, most of the people affected are quite old. Auto accidents, on the other hand, occur "at a time in life when there's a lot of life to be lived. So that if you multiply the number of automobile accidents by the number of years that still could be lived by some wonderful young people," Morison concludes, "one might argue that auto accident prevention was a more important project in which to invest scarce funds."[34]

Another more dramatic example of the political nature of science funding decisions and how "inside" and "outside" considerations can operate against one another is to be found in much of the scientific establishment's attitude toward Department of Defense (DOD) research funding. While individual scientists of note—Philip Morrison, George Kistiakowsky, or Herbert York, for example—protest vigorously against weapons proliferation and the research on which they are based, the National Academy of Sciences and even the Federation of American Scientists either actively solicit DOD support for academic science—funding for funding's sake, it appears—or are silent about the implicit conflict in this matter between the aims of science and society. On the government side, an Office of Research brochure chides "pure" scientists for not applying for funds simply because they do not see "an immediate military application." As the brochure goes on, "We in DOD are interested in all good ideas ... The foremost criterion in selecting a proposal is the *quality* [italics mine] of its scientific content." As one observer concludes wryly: "Official science takes as its main effort the promotion of growth for itself. 'Science for what?' never emerges as an official issue."[35]

If "Science for what?" were to be an official issue, then one might seriously weigh the benefits to be derived from, say, research to make the Post Office more efficient against those from military research or even cancer research. The Post Office, of course, affects all of us in our daily life, but not with the potential to kill; its "inside" scientific quality, for most scientists, would seem nil. Cancer can kill, and its "inside" quality rating is probably middling. Military research can kill, by commission, but its "inside" quality rating for certain fields is very high, i.e., the basic research behind weapons procurement can be fundamental and challenging. How, then, to measure the balance of "inside" and "outside" quality of science, and to decide what to underwrite? Clearly, the decision will be political as well as scientific. It will occur in the context of a political process, which also touches on our deepest and most important value commitments.

In studying science and technology in the manner just outlined, we will have our work cut out for us, for we must understand that politics is both a way of affecting values and then determining whose values are to be implemented. At which point, "Man proposes and God disposes," in the sense that, unless we have knowledge of how secondary and tertiary consequences may occur in unintended fashion, we have little or no chance of having real choice about them. What is worse, the social sciences and humanities offer only very imprecise anticipation of such consequences. To which the answer must be: "on fait ce qu'on peut."

In the matter of indicators of the quality of science, such an answer is, in any event, probably academic. Practitioners of science, whom we have identified as being part of a political process, will be loathe to give up any more control than is necessary over how science is conducted: this is simple political realism, and not evidence of any special self-interestedness on the part of scientists. Political realism, however, also reminds us that there will be continuing pressures, mediated through Congressional representatives, from a public increasingly querulous about what they perceive to be the costs and risks of science and technology as currently presented. Good "inside" leadership, as a practical political matter, must take cognizance of this state of affairs. "Statesmen" of science, or larger-minded "leaders," willingly want to and do recognize the "outside" of the quality of science as well.

V

I have put forward a number of interconnected views: that the "health" or "state" or "quality" of science, as measured by indicators, is necessarily a value-laden concept; that the quality of science, to favor that way of conceptualizing the problem, bears an "inside" and an "outside" image; that quantitative indicators tend to be concerned with the "inside" (though not necessarily so), and while offering essential and valuable understanding, can also lend themselves to pernicious use and can block other modes of understanding; that historical analogies (though also given to suspect usage) and case studies are alternate modes of understanding the quality of science, especially its "outside"; that science should be viewed as a process more than as a result; that the political is an essential part of that process; that the process of science involves fundamental values; and that leadership in that process is a key element, which should be recognized. In short, science and technology are creative enterprises of the human spirit, with all of the limitations and achievements inherent in that state of affairs. The list above is a long one, because I believe it is essential to have a connected or overall view, one which emphasizes a social and historical context, if one wishes to achieve the fullest possible understanding of the quality of science.

Let me conclude with the specific matter of quantitative science indicators, the "preferred" mode of science (à la Kelvin, to which a whole body of scientometricians is dedicated). I would agree that they are indispensable tools for assessing the quality of science, landmarks in a shifting landscape. They can be done well or badly, copying the faults as well as virtues of economic indicators, or transcending the former. They can and will function as elements in the political process. Most dangerously, they can become myths, instead of markers of reality. The burden of this paper is to remind us that science indicators, as any indicators, are social products that must be interpreted, and interpreted in a social context. Finally, and above all, science indicators are to be both constructed and understood with humility and humanity.

Notes

1. Report by the Comptroller General of the United States, *Science Indicators: Improvements Needed in Design, Construction and Interpretation* (U.S. General Accounting Office, 1979), p. 5. The Report drew many of its examples from "Can Science Be Measured?" by Gerald Holton, in *Toward a Metric of Science*, Yehuda Elkana et al., (New York: John Wiley and Sons, Inc., 1978), pp. 39–69.

2. Bruce Mazlish, "Following the Sun," IV *The Wilson Quarterly* 4 (Autumn 1980): 90.

3. *The Health of the Scientific and Technical Enterprise*, An Advisory Panel Report to the Office of Technology Assessment (Washington, DC: U.S. Congress, October 1978), p. 9.

4. Report by the Comptroller General, *op. cit.*, pp. i and 17.

5. Raymond A. Bauer, ed., *Social Indicators* (Cambridge, MA: The MIT Press, 1966), p. viii.

6. Harvey Brooks, "Scientific Concepts and Cultural Change," *Daedalus* (Winter 1965): 67.

7. Derek de Solla Price, "Science Indicators of Quantity and Quality for Fine Tuning of United States Investment in Research in Major Fields of Science and Technology," unpublished manuscript, p. 6.

8. Derek de Solla Price, "Towards a Comprehensive System of Science Indicators," unpublished manuscript, p. 1.

9. Robert K. Merton, *Social Theory and Social Structure* (Glencoe, IL: The Free Press, 1949), especially the Introduction.

10. de Solla Price, "Towards a Comprehensive System . . . ," p. 2.

11. de Solla Price, "Science Indicators . . . ," p. 2.

12. Report by the Comptroller General, *op cit.*, p. i.

13. Lester C. Thurow, "Economics 1977," *Daedalus* (Fall 1977).

14. "First Thoughts on Social Measurement. Commentaries from Twenty Researchers" (Center for Coordination of Research on Social Indicators, SSRC, Washington, DC, no date), pp. 9–10.

15. Bauer, *op. cit.*, p. xiii.

16. Brooks, *op. cit.*, pp. 68 and 67.

17. *Ibid.*, p. 68.

18. Bauer, *op. cit.*, p. 5.

19. *The Measurement of the Attitudes of the U.S. Public Toward Organized Science*

(National Opinion Research Center, University of Chicago, 4 January 1979, prepared for the National Science Foundation).

20. See Bauer, *op. cit.*, p. 31, for this example.

21. *Ibid.*, p. 17.

22. *Ibid.*, p. 1.

23. *Ibid.*, p. 8.

24. Wolfgang Zapf, "International Publications and Private Actors in Social Reporting," *Social Indicators Newsletter,* No. 10 (September 1976): 4.

25. Bruce Mazlish, ed., *The Railroad and The Space Program: An Exploration in Historical Analogy* (Cambridge, MA: The MIT Press, 1965). The definition of a social invention is from p. 11.

26. Report by the Comptroller General, *op. cit.*, p. 5.

27. *Ibid.*, p. 20.

28. Bauer, *op. cit.*, p. 50.

29. *The New Yorker,* 18 February 1980.

30. Bauer, *op. cit.*, p. 5.

31. Leo Marx's remarks were made in discussion at the Harvard-MIT "Indicators of Quality in Science" seminar series, 1 March 1980.

32. John Ziman, "Science Policy and Science Publishing," 3 *London Review of Books* 11 (18 June 1981 to 1 July 1981): 11.

33. For details of the discussion, see James Turner, *Reckoning With the Beast: Animals, Pain, and Humanity in the Victorian Mind* (Baltimore, MD: The Johns Hopkins Press, 1980).

34. Robert S. Morison, presentation to the Harvard-MIT "Indicators of Quality in Science" seminar series, 9 June 1980. Also see Professor Morison's essay "Needs, Leads, and Indicators," this volume.

35. Wil Lepkowski, "The Military Invasion," *The Boston Globe* (21 May 1981): 11.

Industry Evaluation
of Research Quality:
Excerpts from a Seminar

Lewis Branscomb

Editor's Introduction—*On 24 November 1980, Lewis Branscomb of IBM Corporation—who is also the Chairman of the National Science Board for the National Science Foundation—described to participants* in a Harvard-MIT Faculty seminar some of the ways in which American high-technology industry evaluates research quality. Because of his many years of involvement with science policy at the Federal level, Dr. Branscomb set, as a useful "contrasting reference point," the bases for quality evaluation in Federal grants for academic research. The following is an edited version of Dr. Branscomb's seminar and excerpts from the discussion following his presentation.*

Lewis Branscomb Evaluation of scientific research quality is involved not only after the fact—when research is done—but also in resource allocation, in people selection, in project management and review, and in retrospective evaluation. I will discuss today how such matters are viewed in one large, high-technology company—namely, IBM.

Before I do that, I want to provide a contrasting reference point from a point of view that strikes me as infinitely simpler to understand—that is, Federal grants for academic research. I think that the National Science Foundation should allocate grants to research that is investigator-initiated, competitively selected, non-proprietary (therefore published), and of high significance and value. By "significance" I mean the intrinsic values measured by peer review, and I have no other criterion to apply except the opinions of those peers.

* Those present at the seminar included Donald Blackmer, Harvey Brooks, Tom Bryant, Stephen Graubard, John Holmfeld, Gerald Holton, Donald Hornig, Lilli Hornig, Kenkichiro Koizumi, Thomas Kuhn, Marcel La Follette, Leo Marx, Roberta Miller, Robert S. Morison, Alexander Rich, Willis Shapley, Henry Small, Bruce Smith, Christopher Wright, and Dorothy Zinberg.

Value is clearly a multi-dimensional function of an extrinsic nature and it may or may not involve people who characterize themselves as users or beneficiaries. It may, in fact, come entirely from peers. Nevertheless, in that marketplace of research grant opportunity, the satisfaction of needs perceived by those providing the resources are balanced with the perceptions of the investigators.

The probability of success in a grant application is determined in part by the peer evaluation of significance *and* by the success of the applicant in satisfying the peers' expectations of significance (if he can read the minds of his peers). Success is also influenced by the amount of money made available by the allocators of money (which reflects in some measure their view of potential values in an area of activity). The people who put up the resources and manage the grant process may also choose to incorporate in the review process people who happen to be interested in the extrinsic values. So it comes down to the management of that process. Who do you pick to do the reviewing? Who do you put on advisory committees that allocate resources in advance of project applications?

That characterization of intrinsic and extrinsic values does not need to appeal to the language of "basic" and the language of "applied" unless you choose to associate the word "applied" with the notion of extrinsic values. If you do, you will get in trouble, in my opinion. A bubble chamber project may have high extrinsic value to science. A titanium metallurgy project may have high extrinsic value for submarine hulls. The study of line narrowing in lasers might be part of a re-measurement of the Michelson-Morley experiment with a factor of 10^{10} greater accuracy (as was recently done), or it may be a separate project to develop the world's most stable laser, in which case you have to decide whether it is a means to a scientific end and whether that fact makes it basic, or whether it might be a device to be used as a precision oscillator in the optical communications industry. Certainly, the way the work is packaged may influence whether it is perceived as "basic" or "applied." An idea leading to instrumentation may be part of a larger piece of basic research or a separate thing of technological value. The extrinsic values may be values to science, or values to engineering capability, or values to decision-makers choosing among technologies, or values to technologists looking for scientific knowledge to be converted into technology for economic gain. Those are all forms of extrinsic value. . . .

But before I describe research evaluation in a place like IBM, I want to make two comments. One has to do with goals and missions for corporate research. ... The other points out the multi-dimensional character of the things being measured in industry, many of which bear upon decisions about the conduct of research and only one of which is research quality *per se*. ...

The corporate research laboratory may represent anywhere from 1% to 12% of the total R&D investment of the enterprise; at IBM, R&D is about 6½% of sales, and corporate research is roughly 10% of R&D. So approximately 90% of IBM R&D is not corporate research, is therefore mostly "D," and is the driving force that determines how we will define corporate research goals and missions. Obviously, the most important thing to the company is the extent to which it maintains a competitive position technologically, and to what extent this is accomplished within the laboratory responsible for product development.

Indeed, in the "D" we do some unprogrammed, far-reaching, base technology work. ... So, work of a "research" character, for which there is not committed output and which is not part of the planned revenue stream, represents about 15% to 20% of the R&D.

As a company looks at how competitive its technological position is and as it gauges its perceptions of its competitors' technological capability, then it feels heavier pressure as the competition improves. Those pressures are communicated to the research community when the development divisions find themselves hard-pressed to remain self-sufficient in the technology required to create their product. ... the Research Division either feels pressure to do more exploratory development or to do more useful and responsive things, or may volunteer to do more. ...

Needless to say, those perceptions of the company's competitive position are subjective, at least as they relate to technology that has not yet appeared in products. We must guess what the competition will be able to do some years ahead against what we *hope* we can do (not really knowing because we have not yet done it). The dynamics in that perception leave some uncertainty in the self-imposed and externally imposed views of the research organization's mission. It must evaluate and choose in that unclear, real environment.

The second comment I want to make is that we are always measuring a lot of interrelated things at once. (I use "measure" in a hypothetical sense here, because most of these things are not quantifiable...). I will give five examples of things that management tries to track, assess, and sometimes measure, all of which affect research decisions.

First, of course, we want to track the knowledge outputs from the research activity—are they what we hoped and expected to get?

Second, we track the effective use of research outputs, i.e., the strategic and tactical technology inputs to the development organization. So, as we ask what ideas the development organization embraces and where those ideas come from, there is an implied evaluation of the Research Division, even though the Research Division does not have the power to force the development divisions to use their good ideas. They have an obligation to do so in their self-interest, and they work at it.

And, *third*, personnel development for the total needs of the corporation. Often, we look at work more in terms of the attributes of the team leader or the key people involved... than at the documents they produced and the knowledge those documents contained. That is in part because (a) the documents are a pale shadow of the real ideas embodied in the people who did the work and we know that the only really effective technology transfer mechanism is to move the people around or, in any case, to "contaminate" others with the person, not just with his ideas or his documents; and (b)... we often assign people to a project more for the purpose of developing them than generating knowledge....

Fourth, the inevitable pressures for financial efficiency. Those pressures are exerted as much on the Research Division as on everybody else and become part of the measure.

And, *finally*, there is an amorphous category I will call "scientific and technological image".... We do not start research projects for the purpose of achieving that goal, but some research projects are very popular for this purpose and it helps explain why some projects unlikely to be of economic benefit for 10 or 15 years nevertheless have substantial sums of money spent on them.

... The distinction I have tried to draw—between research and development—is one deriving from an overt desire to produce two different environments for operating technical activities, with the expectation that if we make the environments different, then the outputs will be differentiated. It comes about *that* way and not because we look at the work and realize that there are two very, very different categories of work involved, one called research and the other called development, and that they are so different we need two different institutions to do them.

The fundamental fact is that the business is managed in a very micro- (even a pico-) economic way. We begin with a set of business plans aggregated with very large numbers of specific product and technology ideas from the various development and manufacturing divisions. Each division has been allocated part of an overall revenue and profit stream projected to grow appropriately into the future. ... Their business plans must survive review by various skeptical staffs who claim that the results will not be as good as imagined. Finally, when all comes together, we add up a substantial development bill implied to accomplish the whole plan. . . . The engineering, or product development, activity is, on the face of it, thought to be a sufficient activity to generate products within that five-year plan. If we did not care about being in business eight years from now, we would not need to do anything else. IBM created a Research Division because the company does want to be in business in the future. The activities that it takes to insure that long-range future and the independent, informed technical judgment necessary to choose intelligently among technology paths—those two capabilities would simply die if left to the development divisions to accomplish.

Now that makes for some differences in the way we measure. ... I will make the obvious contrasts. At IBM the development environment involves an economic test (at the least, a qualitative one) of every activity—that is, a hypothetical business case is associated with the justification of any development expenditure. We do *not* ask the Research Division to quantify the economic benefits of their activities. Development demands a high degree of dependency on other people's activities. If a development division decides to stop engaging in a project . . . the consequences of their stopping it are interlaced with a lot of other complicated business decisions, whereas the Research Division, which has a low degree of willingness to stop something that looks promising, has a very low degree of dependency on anybody else. The strategic value of results is, we hope, very high in research; it may be low or high on the development side. The mode of utilization of the activity is, of course, different in research—people, publications, and tools. In the development environment, it is processes (that typically are documented) and products (that are only documented once they are committed to manufacturing).

Now, as I begin discussing the characterization of value management, I will first observe that the people at IBM who allocate funds and judge the value of work tend also to be those who supervise the conduct of the work. The best person to measure the value of research is the person whose job it is. If there is conflict, we have a Corporate Technical Committee to help resolve it. If a researcher is infatuated with his own unrealistic notions, there are checks and balances. But, in this respect, IBM is no different from any centrally funded laboratory where a director can measure and then start and stop and reallocate money. The fact that the value judgments are made not by some independent government agency or by peers in distributed locations but by those directing the work, means that the judgments are less independent of the conduct but perhaps better informed. . . .

The second feature is that our peer review is much less formal, but nonetheless real. For example, managers use competition as a very important element in project selection. We spend more time on selection and evaluation in midstream than on retrospective evaluation. . . . Each level of management asks the people to bring the programs they want to do. Of course, everybody brings in a longer list than there is money to support or people to do, but they are encouraged to bring in that longer list. . . . This measurement and evaluation is relatively more open in the Research Division than it would be in a product division. Lots of people are viewers and commentators on that competitive process. For example, the IBM Fellows—about fifty of them, half in research and half in the product divisions—are senior technical people who have a specific charter to put their noses in places where they are not invited and to express their opinions to management all the way up to the Chairman of the corporation. They do so. These people feel immune from retribution and comment quite freely on the projects in which they have no confidence. They represent, therefore, a kind of *de facto* peer review. The research labs, of course, are also our windows on the outside world. The research management needs the outside world's peer review of our insiders to know whether the IBM researchers really know what's going on. We use outside reviewers, for example, in our publication review. . . .

We also spend more time deciding which things to do and how much effort to allocate than trying to evaluate after the fact. There are four reasons why.

(1) The decision process in our environment does permit priority setting. The total resource *is* available to a single decision point, to one person who can delegate that responsibility, and the total resource is relatively inelastic. Our Research Division's budget is determined by the Chief Executive in the presence of the Corporate Management Committee, which will defer to his judgment. . . .

(2) How the Chairman decides is not very obscure; he uses his judgment. Somehow or other, the company has established a Research Division expenditure level of some percentage of all R&D. . . . So the debate takes place over increments or decrements to the established level of effort. The Corporate Technical Committee helps in evaluating those increments and is involved behind the scenes all year long in trading off new ideas against the base. The Research Director knows that although he has a very inelastic amount of money, it nevertheless is very much under his decision control . . . .

(3) Evaluation is a major fact of life throughout the duration of a project. By the time it is completed, a project has been evaluated over and over again, primarily against the expectations at the time it was approved . . . . Of course, the outside world is continually changing and external factors may change prior to the completion of the original plan. Indeed, a project's first phase of activity may determine that it should not be continued. In some respects, if that is why it is not continued, then it may be regarded as a pretty successful project. Negative information can save enormous amounts of money for a division that may be pursuing a goal that is scientifically unrealistic.

(4) Successful project outcomes do tend to be followed by real events. To the extent that it is easier to focus on real events as retrospective measures then we, in some sense, assert that retrospective evaluation is 100% extrinsic. The intrinsic values have either been obtained or not, but in any case are considered immeasurable and of primary concern only to the individuals themselves or to those in the research management worried about the external reputation, recognition, and visibility of their people. . . .

The inverted pyramid of real and opportunity costs, the incentive to focus on exploiting success and not avoiding weakness, is an important perception. Research quality measurement needs a very nonlinear instrument. Let me say that in a different way. More than 90% of our technical activity is involved in product development and base technology. Hundreds of millions of dollars are spent on the tooling for such a manufacturing process. If we are wrong, it is very expensive. In the last ten years, significant sums of money have been invested in capital equipment on processes and so forth that turned out to be wrong. At the time they were being planned, the best technical minds thought they were right. A good research director does not spend most of his time focusing on departments that have poor people in them and are not doing much. If they are not doing much, it only wastes money. On the other hand, if the positive capabilities of the best people (e.g., those who had the potential to invent the Selectric ball) are not exploited, *then* you have *really* lost a multi-million dollar investment. . . .

The top management, then, evaluates research by, first of all, adiabatically changing the base resource as a percent of R&D and then . . . arguing about the increment. . . . The Research Division submits a plan with every single project justified and accompanied by a philosophical document that states what the Division is trying to do and why this is the correct structure for the whole program. So it is possible to debate that program and to challenge the assumptions that underlie portions of it . . . .

Although many companies let development organizations vote on the Research Division by deciding whether to fund the work, we encourage that process as a debate. The user development divisions are specifically asked to review the work of the Research Division, to tell Research what they want them to do (knowing that Research does not have to do it), to appeal a negative decision to me, and if they want, to seek money for the Research Division to fund desired work. General Electric, for example, does more of the latter than we do and, in fact, has a rather quantitative way of scoring projects and of having the user divisions fund some fraction of them. . . . We do try to get divisions to invest in a group of people at a Research Division location, working jointly in a team composed of both groups, and are contemplating putting research people in the development lab locations.

Finally, we have a corporate staff review process, which includes my Corporate Technical Committee. That Committee interrogates the state-of-readiness in the company, including at the research level, vis-à-vis what we perceive as competitive capabilities worldwide. We organize task forces of technical people to study the opportunities and risks of specific programs—for example, on Josephson superconducting technology. Interestingly enough, most of such a study will not be an interrogation of the quality of the research team's work, but it will be an evaluation of the work on competing technologies and whether a success is worth having for x million dollars over y years. . . . We also have a Science Advisory Committee of distinguished outside scientists and other consultants who review work in the laboratories, particularly the proprietary work, precisely for its intellectual and scientific merit as well as for the Committee's perception of its value to the company.

Willis Shapley (American Association for the Advancement of Science) Will you describe further how you divide work between the Research Division and the development divisions? Is it the responsibility of the Research Division or the development divisions to keep track of the advancement of science in the fields with which they are concerned? In other words, is one of the Research Division's functions to keep the development divisions informed of work taking place within the discipline . . . ?

Lewis Branscomb That's a good question. We do not expect the development divisions to be knowledgeable about all the active areas of science which may indeed impact them, because, by and large, they have engineers, not scientists, in those divisions. . . . The Research Division, however, does not accept the responsibility for the purpose of informing the development divisions; they do it for the purpose of ascertaining what activity to undertake in the Research Division.
 . . . if you are talking about outside *technology* activity, then the development divisions are expected to be able to describe the technology assumptions underlying their business plans. If they are starting to work on a computer that will be shipped four years from now, for example, their business plan must contain quantitative assumptions about the performance and price of the competitors' computers at that time. That can only be done by having some notion of the competitors' technology capability, which it is their responsibility to know. . . .

Christopher Wright (Carnegie Institution) How does the corporation see the role of the substance and quality of research in connection with the perceptions of the business you are in? To what extent are these pure flights of the imagination or business judgments of the generalists and to what extent are these judgments made by the scientific and technical people involved?

Lewis Branscomb ... My office and staff were really constructed for the purpose of identifying the technical issues involved in alternative business futures. ... When I took the job, my predecessor, Emanuel Piore, said that the most important legacy he was giving me, and one which I must not lose, is that I am the only officer in the company with no job description.
   ... The Research Division does play a lesser role in debates about future directions. That is, in part, for historical reasons and also due to caution on the part of the research directors. The Research Division is rewarded by sticking to technical matters, in which they can be as radical as they wish and are not subjected to an economic test. ... So they work with the development divisions, but do not often challenge them in their business plans.
   Now there is an exception ... when the Research Division has an idea with commercial significance which they cannot get a development division to take on. ... In that situation, the Research Division can find itself defending its views of the future of the business. Nevertheless ... the Research Division tends usually to be biased towards the science and applied science that underlie the hardware-based technology. ...

Gerald Holton   I see in your presentation a kind of Old Testament, Jehovic view that I am sure you want to modify somewhat—that is, the idea that the Chairman's vision imposes itself on various activities, and that the feedback is either tentative or through someone like you or through the IBM Fellows, and somewhat through the sales. In fact, there *are* certain social values and social goals out there against which your drive is taking place. What is the degree of feedback concerning those social values? Do you measure, do you have openings by which the more-or-less unfocused wishes of society are calculated into your development and research goals?

Lewis Branscomb   First, the Jehovic view relates not to telling people down below what is important for them to do, but to being very supportive of a subjectively justified investment, the details of which the Chairman knows he may not understand. He would not tolerate as soft a justification in any of the ordinary business activities of the company (at least, knowingly tolerate it) as he would in a research venture, because he understands that having the research venture is important to the long-term future. He is prepared to defend a venture so long as he has confidence in the people who are running it. And *that* is the real issue. . . .

Social goals and values do permeate the thinking of some fraction of the individual researchers in the Research Division, and many ideas come from individuals who are so motivated. In fact, some are so strongly expressed that it is extremely difficult for the Research Division not to engage in some purely value-oriented activities. One fellow has a totally different idea about how to do much more productive organic farming, and he insists on thinking about it some part of the time. So, the Research Division does indeed propose things that have that kind of inspiration at their origin. When they fit with rational business justification, then people tend to respond to them—until they get to the point of investment, where it does matter whether it is practical or not.

If the societal values are perceived through the eyes of our customers, and are communicated through the sales force, that is infusion right into the aorta . . . which automatically gets to the heart of the business response.

Editor's Note: The discussion in the seminar then turned back to the question of how the National Science Foundation is structured and to the Foundation's response to the need to support both science and applied science even when the resources may be limited.

Lewis Branscomb In Fall 1979, the National Science Board's attention was directed to the enabling Act of 1950 which established the National Science Foundation, because the Science and Technology Committee in the House of Representatives was giving some thought to examining whether the statute needed to be upgraded or updated. In the course of those discussions, the Board talked about the way in which applied research is treated in the Act . . . . The law set up the National Science Foundation to do basic research, and then it later was amended to provide that if the President issues an appropriate Executive Order, it can also do applied research . . . . So the Foundation has been, for a period of some years, dealing with various versions of the concept of applied research and the NSF role in it. Engineering was a feature of the original Act. The Foundation always supported research in engineering activities, mostly in universities, just as in science; the only real issue is whether or not the magnitude of engineering disciplines ... implies that there should be more budgetary allocation in that field.

The argument on the other side has always been that when the concerns of engineering in this country were not directed to the domestic economy, which was assumed to be in robust health, the real issue was whether the engineering community was in a good position to provide the personnel and the research to achieve the nation's goals in technology (which were mostly government goals). The mission-oriented agencies have always been substantial sponsors of engineering activity. It has been assumed that the Foundation has a special role to lead in basic research, whereas, at least in the applied aspects of engineering, engineers benefit from having real problems to work on and therefore mission agency support is healthy for academic engineering. . . .

So with that background, the National Science Board began discussions that continued through Fall 1979 and Winter and Spring 1980. . . . Engineering and applied science were issues in two kinds of discussions. One had to do with what is now being phrased as the "crisis of engineering," noting that there are two perceptions of the crisis, one of which has to do with engineering as the key activity in industrial performance. There, the engineering industry and the capability of the *profession* are key interests. . . . The other "crisis" in engineering—not totally independent of the first—is a problem for engineering schools. There is now a serious problem of inadequate faculty to teach enough graduate engineers and to produce people at the doctorate level in engineering who can expand the faculty to meet the very heavy demand for all engineers. It is similar to the situation in computer science, where industry hires people at the Master's level, before they get the Ph.D., and faculty salaries are not always comparable to industrial salaries in crucial fields. Furthermore, students in many fields of engineering do not have access to the newest equipment, and therefore do not acquire the design skills and experiences that leading industry represents, and are not as well-trained for an industrial career as they might be. . . .

The academic engineers are very interested in engineering as a *discipline* as well as a profession, and they point to a fact that I believe to be true, namely that engineering is a collection of disciplines ranging from what in other disciplines we would refer to as very basic to very applied. There are very mathematical, abstract, theoretical branches of engineering; there are straightforward, practical aspects of engineering. Within the field of engineering, a spectrum of "basic" to "applied" co-exists in each of the standard disciplinary names of engineering, and there is a dependency of industry on the university base, and new problems in that university base. Therefore, the engineering problem is perceived by many people as calling for resources that do not now exist, resources aimed at strengthening the level of technical sophistication, the Ph.D. level of proficiency, and the really modern technologies at the forefront. Some people in industry believe that that is exactly the part of the engineering problem on which, in its self-interest, the government should focus.

The other discussion had to do with great difficulties in the language ("basic" and "applied"). ... The Board has tried to put down in writing its point of view on these kinds of questions in a variety of contexts, but there are hidden agendas. ... Indeed, I think the National Science Foundation is in a very poor position to deal effectively with any substantial amount of research for which the research justification consists of a perception of a national social problem, the belief that research is an important factor in solving social problems, and the expectation that the research will ultimately deliver the social benefit. When the government asks the researchers to work on a problem of the government's choosing, that is mission-oriented research. If it is done by the NSF, then it is undertaken in an agency that does not have the mission to deliver the end product to the consumer. ...

Many people believe that if the Foundation acknowledges that "applied research" is a significant area of activity, then it will quickly become vulnerable to pressure to solve all the country's problems ... . The best defense against that is to say "We don't do applied research, we fund basic research." ... But, having taken that position, how in the world are you going to explain in plain English to any citizen, let alone to any Congressman, why we should spend a billion dollars a year to improve our culture—however important science is in demonstrating man's image of his own glory and destiny?

In many peoples' view, most basic research ... has substantial, definable value. When you are trying to explain the benefits of the relationship between science and the well-being of mankind, it is a mistake not to be able to aggregate the external values perceptible throughout the Foundation's activities. So the second feature of the discussion that took place in June was the notion that if we move away from debating how to constrain and manage those kinds of applied research that do or do not make sense—all the while claiming that, in any case, the primary mission is basic research and applied research will always take second seat—then we will continue to suffer difficulties.

There *is* an alternative. Instead of having applied research segregated in a box called the Engineering and Applied Research Directorate, which conveys the impression to the engineering community (as well as to others) that applied research is a kind of engineering and that all engineering is applied (neither of which is the case), we could distribute throughout the Foundation the responsibility for dealing with the appropriate kinds of research of high intrinsic and extrinsic value. We could use as a criterion for choice the notion that what *should* be supported is investigator-initiated, non-proprietary published work of high significance (i.e., intrinsic value) and importance (i.e., extrinsic value) as competitively judged by the existing peer review mechanism. Now, neither of those statements, on the face of it, has anything to do with organization. One simply says that there are problems to fix in engineering; the other says that there is a more sensible way to manage what many people call applied research without making the rigid distinction between "basic" and "applied," and in a way that will protect the inherent values of what we mean by "basic."

The Public and Science Policy*

Kenneth Prewitt

Kenneth Boulding, speaking to the American Association for the Advancement of Science, observed that science occupies a sociological niche, the boundaries of which are

> ... very largely determined by the image of science in the minds of nonscientists, especially those who make decisions about budgets, whether in government, education, or industry. If this image changes unfavorably, the niche will begin to close ... The scientific community, therefore, should be deeply concerned about the images of science that lie outside it and even those that lie within it, for the probability of adverse changes in these images is at least large enough so that ignorance about them would be unwise.[1]

Boulding's observation, if extended, serves as text for the present article. If the boundaries of science's "niche" are fixed by images held by its funders in government, education, and industry, these funders in turn occupy their own niches, whose boundaries are fixed by the images of citizens who provide the funds eventually dispensed to science. Government funds for science are tax dollars, university funds for science are tuition payments and endowment gifts, industrial funds for science are profits earned on sales. The images of science held by those whose dollars support the government, universities, and industry matter as much (or more) as the images of the immediate funders themselves.

Boulding's recommendation instructs us to search deep into nonscientists' images of science. Such investigation must reach out even to where the specifically "scientific content" of the images has been lost, or has become absorbed into citizens' views about privacy or safety or national prestige. These views eventually affect larger sci-

---

*Portions of this paper were initially presented at the third meeting of the Harvard-MIT faculty seminar series on "Indicators of Quality in Science," April 1980. I am indebted to the participants of the seminar for wide-ranging critical commentary.—KP

ence policy questions—perhaps even more than narrowly construed attitudes toward science do. These views can, without doubt, affect adversely as well as favorably the image of science held by those who more immediately control funding and allocation decisions for science.

Much of the discussion about the public images of science has focused on what the public knows, or should know, about science—a topic variously labeled as the public understanding of science, or scientific literacy, or public education about science. This general topic has been the subject of special issues of major intellectual and professional journals,[2] the focus of the 1981 NEH Jefferson Lectures,[3] of concern to those who are building a science indicator base for the society,[4] and the rationale for the establishment of various specialized information and education programs.[5] Although observers vary in what they describe and recommend, the general conclusion seems to be that scientific literacy is too low, and something should be done about it.[6] In much of the analysis, the levels of scientific literacy or public understanding are explicitly linked with what is perceived to be a growing "anti-science" mood in the nation. If the public only understood us better, many scientists claim, they would love us more and demonstrate this affection by increased support of science.[7]

The literacy issue can, of course, also be stood on its head—which is perhaps what Boulding is recommending. Instead of asking what the public knows or should know about science, we might ask what the scientists know or should know about the public. There is some evidence, in fact, that the public knows more about science than scientists know about the general public. With the exception of social scientists whose research focuses on public belief systems, the scientific community has often shown itself to be ignorant about the way in which public attitudes are formed and become expressed through the political participation system. I can provide some anecdotal evidence. My colleagues and I were under contract to the National Science Foundation to design a survey questionnaire for the public attitudes toward science chapter of *Science Indicators 1980*. We recommended that the public be asked not only general questions about science and technology, but also focused questions about specific controversial issues in science and technology. We provided questions on three such controversies: food additives, nuclear plant siting, and Recombinant DNA, or genetic engineering. In responding to the draft questionnaire, National Science Board members and Foundation staff asked us to consider three alternatives: nitrogen fixation, black holes, and plate tectonics. Only persons unfamiliar with public opinion would imagine that anything of importance could be learned by asking the mass public questions on such highly specialized, low visibility issues.[8]

Scientists' inadequate understanding of the public has direct bearing on the concerns that Boulding would urge upon the scientific community. How can we deal with the image of science among non-scientists without some prior understanding of how non-scientists come to hold and act upon their images? Indeed, how can we react to these images if we do not even have them correct? A few years ago, for example, there was great concern about the "declining level of public confidence in science." The National Science Board (NSB) highlighted this issue in its 1976 annual report, published as *Science at the Bicentennial*.[9] This report described the responses from 900 people who were active in the administration and conduct of research in the United States and who worked in one of four sectors (universities, industry, Federal laboratories, and independent research institutes). In his letter of transmittal to President Ford, NSB Chairman, Norman Hackerman summarized the report's conclusions by noting that the "[g]reatest concern [of the respondents] centered upon dependability of funding for research, the vitality of the research system, freedom in research choices, and [public] attitudes toward science and technology." According to the report, respondents from all sectors shared the belief that the public's confidence in the scientific establishment had declined to dangerously low levels. The section on these conclusions, "Confidence in Science and Technology," opens with a quote from Stanley J. Lawwill, president of Analytic Services, Inc. (an R&D center sponsored by the Department of Defense):

The number one problem which I see facing fundamental (long-term, basic) research in the near future is the poor, and deteriorating, National attitude toward science and technology. Until this trend is reversed, I see little prospect for the United States regaining the dominant position it once held in the discovery of scientific knowledge and in the development and application of technology.

Another respondent, A. S. Gregory, Director of Central Research and Development at Weyerhaeuser Company, concurred:

The number one problem is society's attitude regarding importance of scientific and technological advances . . . Recently, it has become a popular game of the uninformed to state that we have all the science and technology we need and that many of our current problems stem from past technological advances. A significant sector of society does not seem to realize that many of the things that give us our preferred quality of life are possible because of technology.

These scientists' images of the public were similar to those which appeared in a wide variety of publications, but were not supported by the evidence at hand. The data most commonly cited were a series of trend lines showing the percentage of the public which had "great confidence" in the people running each of nine American institutions, one of which was scientific institutions. These trend lines span the period from 1966 to the present, and therefore reflect general public disillusionment prompted by the Vietnam War, the forced resignation of a U.S. President, and public disclosure of corruption and illegality in the very highest political circles. Not surprisingly, the data show declines in levels of public confidence. Given the events of this period, we would have had cause to wonder about either the intelligence of the American public or the reliability of survey data if the levels *had not* declined.

No institution, including science, was immune to this shift in public attitudes. This fact brings out an important point about public opinion: levels of public confidence tend to increase or decrease for all institutions together, rather than for each separately. As one analyst of these data concludes, "These results suggest that public confidence in an institution is less a function of its own behavior than of larger forces external to it and beyond its control."[10] Within the overall decline of public confidence in American institutions, science, in fact, fared comparatively well during the period 1966 to the present. Of the nine institutional sectors measured, science stood fourth in 1966 and had increased its relative standing to second by 1980 (medicine still being first, of course, resting on great advances in the biological sciences during this period). Whatever departure science has shown relative to the modal trend, observes Mazur, "has been positive, moving up in relative confidence . . . and enjoying nearly as high an absolute confidence level now as it had in the last decade,"[11] a conclusion he is not alone in making.[12]

The "confidence in science" trend lines are not the only data suggesting that the National Science Board correspondents may have lifted isolated facts out of context and overinterpreted them. For instance, over the last two decades, there has been an erosion of public confidence in inevitable progress; however, science and technology have been less affected by that erosion than other sectors of American society. Many prominent people have declared that the prestige of the United States in the world is declining; however, the public believes that American science and technology continue to contribute more than other sectors to the prestige we do enjoy. There is certainly a growing public skepticism about society's capacity to supply itself with the material and organizational resources necessary to sustain continuous increases in living standards; however, in the public mind it is our scientific strength and technological know-how that will provide such tools as are available to us. Pressure for participatory democracy has obviously grown; however, there remains in this society a clear deference to scientific expertise and competence on many complex issues.[13]

Across a series of measures and issues, therefore, there have been changes in public opinion and American self-conceptions over the last two decades, and science and technology have not been immune to these broad trends. But within the trends, public respect for science and technology is firmer and better-anchored than perhaps is the case for any other institutional sector and its related cultural values.

The purpose of this extended example is not to ridicule *Science at the Bicentennial* or the journalistic literature which the report reflects. Rather, I am simply attempting to remind the scientific community that it cannot follow Boulding's advice, cannot concern itself with the images of science beyond its borders, unless it seriously (one is tempted to say scientifically) inquires into how those images are formed, change, and become the basis for political activity. The remainder of this article comments on a few of the many issues germane to such inquiry.

Issue Specialization in Public Opinion

Advanced industrialization leads to specialization in many domains of society: in the division of labor, in the activities of voluntary associations, in the organization of politics, in religious belief systems, in public and private entertainment, in publishing and the media, and, of course, in science and technology. Specialization even affects public opinion, taking the term "public opinion" to mean those views and images held by the general public which political leaders find prudent, maybe even necessary, to heed as they shape and implement public policies. It should not surprise us that specialization occurs even within that domain of public life we refer to as public opinion. Not every citizen is equally knowledgeable and interested in all aspects of public policy. Many citizens are not much interested in any aspects of public policy, giving their attention primarily to sports or fashion or movies. Other citizens "specialize" in a narrow area of public policy, such as business policy, education policy, or labor policy. Political scientists have coined the term "issue specialization" to describe this tendency for members of the public to concentrate on some selected number of public issues rather than try to maintain interest and knowledge across the complete spectrum of policy domains.[14] Those citizens who follow the developments within a particular domain have also been called the attentive public for, say, foreign policy or civil rights policy or, as we shall see, science policy.

The notion of an "attentive public" was first advanced by Gabriel Almond in his 1950 study *The American People and Foreign Policy*.[15] Almond noted that the majority of a population was unlikely to sustain interest in foreign policy matters, becoming interested during periods of war or the activation of the military draft, but then losing interest when international crises subsided. A small part of the population, however, would tend to monitor foreign policy issues more continuously. This "attentive public" is self-selected; it organizes itself in world affairs councils or foreign policy associations; it communicates through specialized journals such as *Foreign Affairs, The Nation,* and *Foreign Policy*. Since the publication of Almond's work, other scholars have provided a rich empirical description of the attentive public for foreign policy.[16]

Other attentive publics have received less research attention, but impressionistic evidence indicates that they exist for, at least, civil rights policy (since the 1960s), agricultural policy, business and taxation policies, and (again since the 1960s) policies that affect the status of women in society. There is the possibility that the Moral Majority will establish an attentive public for the cluster of social issues with which it is concerned (e.g., school prayers, abortion, creation theory).

We know very little about the formation, replenishment, and perhaps decay of attentive publics, what we might call their "life-cycle."

The absence of empirical work on attentive publics other than the one for foreign policy has prevented scholars from answering questions about overlap and interactions among attentive publics, although the issue-specialization notion suggests limits on the number of attentive publics to which a citizen might belong.

My present interest in issue specialization and attentive publics concerns what these concepts may reveal about the relation of the public to science policy. It is becoming apparent that there is an attentive public for science and technology. In his 1950 work, Almond had suggested that individuals belong to an attentive public (for foreign policy) if they 1) display a high level of interest in the topic, 2) are generally informed about the topic, and 3) are committed to a pattern of information acquisition which assures a continuing high level of knowledge about the topic. Almond's attentives are interested and informed about foreign policy, and subscribe to specialized journals and belong to appropriate organizations which help them to maintain this interest and information.

By these standards, the attentive public for science extends well beyond the scientific community itself. The size of this attentive public depends, obviously, on the threshold set for levels of interest, knowledge, and sustained information acquisition. And, for present purposes, it is not necessary to work our way through the various criteria, and correlations among the criteria, which can be used to circumscribe the attentive public. A 1979 study by myself and colleagues concluded that approximately 18% of the American public belong to an "attentive public for science."[17] If the criteria were loosened, this proportion would increase, although not by much or the concept would lose meaning. If the criteria were made more restrictive, the proportion might drop to about 10%. Even under the more restrictive criteria, the attentive public includes a large number of non-scientists—such as persons who watch "Nova" or the National Geographic television specials, who subscribe to *Science 82* or *Psychology Today*, who frequent natural history and science and technology museums, and who regularly read or listen to science news in the general media.

Dividing the general population into the attentives and the non- (or less) attentives leads to an important observation about the public and science policy. The attentives will relate to science policy issues quite differently than the non-attentives, indeed so differently that we require separate analytic models for the two parts of the public.

The Attentive Public and Science Policy

The term "science policy" can and does mean many things. In this paper, I have in mind government decisions (or restraint from decisions) that affect where, how, and with what resources science is practiced in the United States. Science policy, in this limited sense, includes the level and allocation of public funds available for research and training, as well as publicly imposed incentive/disincentive arrangements, such as tax credits, that affect private support for science. It includes such issues as the degree of control exercised over science by non-scientific agencies or procedures, the protection of human subjects regulations, safety standards in scientific laboratories, and cost-accounting practices applied to scientific grants and contracts. Under this general heading would certainly be included restrictions on the kind of science practiced, as in the attempt to restrict Recombinant DNA research or inquiry into hypothesized genetic differences in IQ. Science policy also includes publicly imposed demands on science: e.g., put a man on the moon, cure cancer, increase economic productivity. These demands more often express themselves through technology rather than directly on science, but the linkage is too tight to presume that science cannot be re-directed by the demands.

Science policy, even under the restrictive definition offered, ranges from the narrow and specific to the broad and inclusive. The day-to-day decisions actually taken by science policy-makers mostly resolve fairly narrow questions; more often than not they involve shifting resources at the margin. Although these funding shifts are consequential, usually casting a long shadow, they do not resolve the kind of policy question that captures public attention, even the attention of an "attentive public." Toward the broader end of the science policy issues, however, are found questions that might attract an attentive public. Should science funds be primarily allocated through a peer review process or according to programmatically set goals? How much of the nation's research effort should be conducted in universities, in Federal laboratories, in industry and other for-profit institutions?

Although we do not yet know much about the attentive public for science, and how it might relate to science policy issues, three points can be made with some confidence. *First,* compared to the attentive public for foreign policy, the attentive public for science policy is less internally organized. The foreign policy organizations and journals that provide opportunities for dialogue have been in existence for some time. Set labels categorize foreign policy positions, and give coherence to discussion and debate: e.g., hawks and doves, isolationists and internationalists, the bipolar vs. the multipolar strategy. There are also close and continuing links between the attentive public and the policy-makers at, for example, the State Department, the National Security Council, and the Department of Defense.

Similar structures have yet to evolve with respect to science policy. In part, this is because it takes time for internal structures and organizations to evolve, but science policy as a phenomenon is also less fixed in the public mind. There are no generalized world-views—such as, for example, the containment doctrine or détente in foreign policy—that could provide orientation for how even to begin to think broadly about science policy. For the most part, public discussions of science policy have been about funding levels and peer review systems.

I believe that fairly rapid developments are occurring which will lead to more structure and coherence in the attentive public's discussion of science policy; the attentive public is becoming more internally organized and policy questions will become framed in terms of orienting concepts. Indeed, greater internal structure and coherence of this group is, in my judgment, more likely than its additional expansion. If issue specialization is characteristic of public opinion, the attentiveness in any given policy domain is limited by competition from other domains. If the attentive public is unlikely to expand, it is likely to develop more internal organizational structure and greater coherence in how it reflects upon science policy issues. Already there is evidence that the attentive public is capable of forming reasonably well-informed views about policy questions.

This observation leads to the *second* point. Do policy views of the attentive public agree with or differ from views of scientists themselves? Such evidence as is available (and it is quite limited) suggests that, at least on some policy issues, the attentive public and scientists tend to see things in similar ways.

Starting in June 1976 and concluding in May 1978, the National Science Board sponsored six regional forums intended to "facilitate the participation of members of the public in the formulation, development, and conduct of the National Science Foundation's programs, policies and priorities." The forums, and the language just quoted, were directives of Congress, as expressed in the NSF Authorization Act for Fiscal Year 1976. The Board referred to the forums as "an experiment with public participation in science policy formulation."[18]

The forums, held in six cities around the country, attracted persons who can reasonably be classified as part of the attentive public for science. They also attracted a fair number of practicing scientists. As part of the survey designed to evaluate the forums, a few science policy questions were asked of all participants. Approximately 1,000 respondents were asked if they would alter the program emphasis of the Foundation, what role they would assign to scientists (in comparison to public interest groups, private business, Federal agencies, state and local government, the Congress, unions) in establishing basic science priorities, whether they favored preferential award criteria for women and minorities, and whether they favored increased industrial conduct of basic research. Table 1 shows that the pattern of responses between the scientists and non-scientists across these policy issues is remarkably similar.[19]

Table 1
Policy Preferences of Scientists and Non-Scientists Attending National Science Board Regional Forums

	Scientists*	Non-Scientists (Attentive Public for Science)
Would Change NSF Program Emphasis	55%	55%
Would Assign 'Major' Role to Scientists in Establishing Basic Research Priorities	62%	50%
Favors Increased Industrial Conduct of Basic Research	43%	56%
Favors Preferential Award Criteria for Women	23%	23%
Favors Preferential Award Criteria for Minorities	27%	22%

* The number of respondents varies somewhat from question to question because not all questions were asked in every forum. The number of scientists answering each of the five questions, in the order listed in the table, is: 440, 433, 311, 326, 322. The number of non-scientists answering each of the five questions, in the order listed in the table, is: 532, 530, 386, 400, 404.

Although these data hardly constitute a definitive test, they are consistent with the presumption that the views within the attentive public for science are in reasonable accord with scientists' views on policy issues. This does not imply, of course, that differences do not exist, nor should similar structure across the two groups be confused with unanimity of policy views. Table 1 itself suggests otherwise. For example, both the scientists and the non-scientists split about 50/50 on whether they would change program emphasis at the National Science Foundation; and each group splits about 75/25 on the issue of preferential treatment for women and minorities. It will take far more sophisticated data than currently available to work out the causal patterns, to determine whether the differences within the attentive public reflect those within the scientific community because scientists are providing opinion leadership, or whether the differences among scientists reflect splits in the attentive public because that public serves as a pressure group on science, or whether these are independent phenomena.

What is striking about the participants at the regional forums is the similarity of policy views between the scientists and the non-scientists. If additional studies confirm this finding, then the relation between the public and science policy is far different than the image of it which many scientists appear to have.

This leads to our *third* and final observation about the attentive public and science policy. The "attentive public" was constructed without reference to how its members evaluated science and technology issues. The defining criteria of attentiveness are interest, knowledge, and information seeking, criteria which could as easily apply to a skeptic as a believer, a critic as a friend. There is no *a priori* reason to expect a negative or a positive correlation between level of attentiveness and degree of confidence in or appreciation of science. Although the empirical work is just now getting underway, early analysis suggests that attentives are consistently more favorably disposed to science than the rest of the public. This pattern holds across a series of measures, including expectations toward future scientific accomplishments, a tendency to see more benefits than risks associated with fields of scientific inquiry, a reluctance to impose limitations on science, a view that science and technology are responsible for improving the quality of life, and so forth.[20]

The observation that attentives are generally more positive toward science than non-attentives need not imply that non-attentives hold negative views. On many measures, both the attentive public and the non-attentive public express quite favorable judgments about science, but in nearly every instance the attentive public is more favorable than the non-attentive. How can this finding be reconciled with the view, widely expressed, that there is growing public restiveness with science and technology? Persons who advance this claim are not without concrete examples. They point to the debate over Recombinant DNA research, with its implications for politically established controls over research—a debate that has now died out without the establishment of restrictive controls. They point to demonstrations against the siting of nuclear plants and public controversy over radioactive waste disposal as illustrative, perhaps, more of a growing fear of dangers associated with nuclear technology than of an anti-science attitude. There is also the widening public discussion of ethical and moral issues in science, such as privacy, protection of human subjects, and biomedical ethics. Some of these discussions broaden the ethical issues by raising the specter of subversive knowledge that might undermine valued belief systems, as in the renewed debate in this country over the teaching of evolutionary theory in public schools. As Weingart comments elsewhere in this book: "Public criticism of science and technology may not be new, but its character is changing in breadth, intensity, and focus, one example being that the target of public criticism has moved from the application of science to the research itself."[21]

Such claims obviously contain some truth, but I believe that the idea of an "anti-science mobilized public" is easily exaggerated. More probably, two somewhat contradictory developments are at work simultaneously—one, the establishment and growth of an attentive public for science; the other, the more issue-specific agitation on various science and technology matters. This latter development increases public awareness of science through such events as public referendums, citizen-action groups, and citizen- (or public official-) initiated demands for the discussion of controversial issues such as *in vitro* fertilization or Recombinant DNA research. Because much of this agitation is issue-specific, public involvement is unpredictable and ad hoc—a situation hardly reassuring to scientists who trade in predictability and pattern.

This development must be seen against the backdrop of the growing institutionalization of an attentive public for science, one well-informed and favorably disposed toward science. For certain, within an attentive public there is room for doubt about science and there is concern with the ethical and political questions raised but not answered by science. Our picture of the attentive public is yet too incomplete to know how much it incorporates the range of protest activities causing such alarm in the scientific community. No doubt there is some of this. But I believe that subsequent research will reveal the growth and institutionalization of the attentive public to be generally independent of the agitation and protest associated with, for instance, the anti-fluoridation politics of the 1950s or the current anti-evolution campaign.[22]

Such a model is not intended to dismiss the protest politics, but it does place public attention to science within the current context of an upsurge in participatory politics, the targets for which vary across a wide range of social, economic, *and* science-technology issues. That an increasing share of targets are science-technology issues derives less from "anti-science" public attitudes than from the less dramatic circumstance that many more science and technology questions are now on the public agenda.

The General Public and Science Policy

Although the general public does not pay much specific attention to science policy matters, there are many ways in which public attitudes and images bear on science policy. The assumption that public opinion affects science policy is implicit in the definition of science policy as something government does to the practice of science in this society. Unless one shuts public opinion completely out of the governing process (an argument advanced by only a small number of political scientists), it is true by definition that public opinion will affect science policy. To understand how this might occur requires a search for indirect and quite extended linkages. Many aspects of public opinion important for science policy are not themselves attitudes toward science. They are not levels of scientific literacy or trends in confidence in science or risk/benefit calculations about scientific discoveries. They are aspects of public opinion affecting broad public policies in a manner that has spillover effects on science policy.

This is most easily seen in government spending levels and priorities. For example, public support for increased defense spending and corresponding decreases in social welfare programs will lead to shifts in funding toward mathematics, physics, and engineering and away from large-scale social experiments or multi-wave social surveys. These funding shifts alter support levels across the sciences for reasons wholly unrelated to judgments—whether by the public, by an informed subset of it, or by science policy-makers—about developments within the affected sciences. An earlier set of decisions, in the 1960s, promoted intellectual and institutional developments in the social sciences (policy studies, evaluation research) because there was public support for the Great Society legislation, not because the sciences were demanding, or even ready for, the responsibilities assigned to them.

The circuitous route by which general public opinion impinges on science policy is true not only of funding levels and allocation decisions. It can also be seen, for example, in the policies guiding the international exchange of scholars and scientific ideas. Something narrowly called "science policy" may affirm the importance of an unfettered flow of scientists and scientific findings across national borders, but the actual flow is determined by foreign policies and their public support. Geologists may have wanted soil samples; geneticists plant specimens, historians archival material, and anthropologists village observations from the People's Republic of China in the 1950s or 1960s, but these scientific opportunities were not presented until political leaders (with, it turned out, public support in both countries) normalized political relations between the United States and the People's Republic of China.

This general point can be extended in many directions. There are public attitudes toward privacy, which become translated through very indirect chains into scientific policies on informed consent and the protection of human subjects. There are public views about the treatment of animals (and anticruelty laws) which influence how scientists use experimental animals. There is public support for such basic constitutional principles as freedom of speech, which forms the public context within which debates occur about the freedom to teach even "alien" scientific doctrine. There are, of course, public judgments about the appropriate role of the public sector, and the limits to that role; these judgments, again very indirectly, work their way through policies that eventually affect how much U.S. science will be conducted in Federally supported laboratories and universities, how much in private non-profit institutions, and how much in the commercial sector.[23]

The obviousness of the point I am making should not be allowed to disguise its complexity. What linkages and political processes transmit general public opinion to science policy, when that public opinion is not explicitly focused on science or science policy questions? These are complex and dimly understood processes. That they exist can hardly be doubted. That they merit study is self-evident—if we are to have a scientific understanding of the relation between public opinion and science policy. To restrict our studies to "public attitudes toward science" or to "scientific literacy" is to carve out a very small piece of how science policy is made in a democracy. This point is emphasized in writings on the American public and science policy which use the metaphor "renegotiating the contract between science and society." The point that general public opinion impinges upon science policy indirectly, but not unimportantly, puts this metaphor in further perspective.

The scientific culture is based on notions such as peer control and professional autonomy. Science is guided by self-produced and internalized standards. As a profession, science regulates its own entry and exit criteria, determines its own rewards and penalties.

The democratic political culture starts from a contrasting set of assumptions. Instead of peer control and autonomy, there is public control and accountability; instead of internalized standards, there is public scrutiny, instead of self-regulation and self-evaluation, there are checks and balances, external regulations, and publicly produced evaluations. The political theory of the Constitution assumes the broadest-based policy direction, governed by electoral accountability. The public trust must be continually earned.

These deep themes in American political culture cannot but affect any sector of American life which is public—that is, which takes from the public treasury, which seeks public esteem, which claims public significance, which says it acts in or for the public interest. As science and technology have pervaded the public agenda, then the deeply held political values of democratic accountability and public scrutiny have naturally and inevitably impinged on science policy. Demands for observable benefits from public investment in science increase. And there are expectations that science should be subjected to the same principles of regulation and accountability as govern other important sectors of public life. It is too late for science to call for a "separation of science and the state." It is too late for science to evade the consequences of the participatory democratic culture in which it is practiced and from which it draws support.

This, it seems to me, is the proper context within which to consider the metaphor about contract renegotiation. The contract is not being renegotiated because the general public is "anti-science" or because an attentive public is mobilized against science. It is being renegotiated because science is of public consequence. A public that does not often or deeply think about science nevertheless does instinctively affirm the political theory of the U.S. Constitution.

On what terms will the contract be renegotiated? It would take another essay to begin to answer that question thoroughly, but I believe that the contract can be negotiated on terms acceptable and even favorable to science, and that the attentive public will play a critical role in the renegotiation, just as the attentive public for foreign policy plays a critical role in that policy sector. The general public, in its own way, will also matter, but less in specific science attitudes than in the broader currents of public opinion, including the play of democratic values in that opinion.

This renegotiation will obviously be more difficult and will proceed on less favorable terms if the attentive public is mobilized around antiscience themes and the general public is losing its confidence and respect for science. But the renegotiation will proceed on the very worst of terms if the latter is not the case and yet scientists and science policy-makers act as if they believe it to be.

Notes

1. Kenneth E. Boulding, "Science: Our Common Heritage," 207 *Science* 4433 (22 February 1980): 832–833.

2. See, for example, the Spring 1978 issue of *Daedalus* ("The Limits of Scientific Inquiry").

3. Gerald Holton, "Where Is Science Taking Us?," The Jefferson Lecture delivered in Washington, DC, and Boston, MA, 11 and 13 May 1981.

4. Jon D. Miller, Kenneth Prewitt, and Robert Pearson, "The Attitudes of the U.S. Public Toward Science and Technology," (Chicago: National Opinion Research Center, University of Chicago, 1980), report prepared for the National Science Foundation's science indicator series.

5. For example, The Scientists' Institute for Public Information.

6. For a contrary view, see Leon E. Trachtman, "The Public Understanding

of Science Effort: A Critique," 6 *Science, Technology, & Human Values* 36 (Summer 1981): 10–15.

7. See, for instance, Bruce Smith, "A New Science Policy in the United States," XI *Minerva* 2 (April 1973): 162–174. Smith writes that the expansion of Federal support for science in the 1960s "was influenced by a widespread confidence in science and technology. The decline which set it in the late 1960s was, in part at least, a consequence of the gradual diminution of the public appreciation of technology" (p. 163).

8. The three issues selected were food additives, nuclear plant siting, and space exploration. See Miller, Prewitt, and Pearson, *op. cit.*, p. 90; and *Science Indicators 1980*, p. 169. The latter is the Report of the National Science Board, published in December 1981.

9. National Science Board, National Science Foundation, *Science at the Bicentennial: A Report from the Research Community* (Washington, DC: U.S. Government Printing Office, 1976).

10. Alan Mazur, "Commentary: Opinion Poll Measurement of American Confidence in Science," 6 *Science, Technology, & Human Values* 36 (Summer 1981): 17.

11. *Ibid.*

12. Clyde Nunn, "Is There a Crisis of Confidence in Science," 198 *Science* (9 December 1977): 995; and Barbara J. Culliton, "Science's Restive Public," 197 *Daedalus* (Summer 1978): 147–156.

13. These conclusions are drawn primarily from Miller, Prewitt, and Pearson, *op. cit.*

14. Jon D. Miller has suggested the utility of the concept "issue specialization" in several working papers. See Miller, "Selective Attentiveness: A Conceptual Framework for Understanding Public Attitudes Toward Organized Science," a paper presented to the 1978 Annual Meeting of the Society for the Social Study of Science, Bloomington, IN (November 1978) and Jon D. Miller and Kenneth Prewitt, "The Attentive Public for Science Policy: A Case Study in Issue Specialization," presented to the 1981 Annual Meeting of the American Association for the Advancement of Science, Toronto, (January 1981).

15. Gabriel Almond, *The American People and Foreign Policy* (New York: Harcourt, Brace, and Co., 1950).

16. B. C. Cohen, *The Public's Impact on Foreign Policy* (Boston: Little, Brown, and Co., 1973); J. E. Mueller, *War, Presidents, and Public Opinion* (New York: John Wiley & Sons, Inc., 1973).

17. Miller, Prewitt, and Pearson, *op. cit.*, p. 43; and *Science Indicators 1980*, p. 177.

18. "Regional Forums of the National Science Board: An Experiment with Public Participation in Science Policy Formulation," report of the National Science Board, prepared by Ray Bye, undated.

19. Table is taken from Robert Pearson and Kenneth Prewitt, "The Attentive and Mobilized Public for Organized Science: A Case Study," presented at the 1979 Annual Meeting of the American Association for the Advancement of Science, Houston (3–8 January 1979).

20. Miller, Prewitt, and Pearson, *op. cit.*

21. Peter Weingart, "The Social Assessment of Science, or the De-Institutionalization of the Scientific Profession," this volume.

22. A recent contribution of Allan Mazur, however, suggests that public protest against intrusive or potentially unsafe technologies is perhaps becoming institutionalized through the environmental movement. The overlap of this structure with what I have called the attentive public for science is yet to be sorted out. See Mazur, *The Dynamics of Technical Controversy* (Washington, DC: Communications Press, Inc., 1981).

23. *Ibid.* In his analysis of public controversies over technical issues, Mazur comes to a conclusion similar to the general argument advanced here. To fully understand the fluoridation controversy, for example, requires seeing it in the context of McCarthyism of the 1950s. The technical controversies studied by Mazur all have a "political and nonscientific" context that bears on how an issue becomes controversial and how it is resolved.

Changing Public Attitudes to Science and the Quality of Life: Excerpts from a Seminar

Daniel Yankelovich

Editor's Introduction—*On 8 December 1980, Daniel Yankelovich outlined to participants* in a Harvard-MIT faculty seminar his analysis of the changes that are currently taking place in American values and attitudes, changes that could have profound political implications for science and technology. The research project he describes has been undertaken by the Public Agenda Foundation, a nonprofit nonpartisan organization, under subcontract to NSF Grant SRS-8007378 to Harvard University.*

Daniel Yankelovich Using ordinary polling methods to gauge the judgments people make about how science and technology affect their lives presupposes a false model of public attitudes and opinions. It presupposes that such judgments are held by individuals in an isolated form and are not part of a group process. And it also presupposes that they are instantly accessible, "top of the mind" responses—the same type of judgments made when people are asked, "What brand of toothpaste do you prefer?" or "If the election were held tomorrow, who would you vote for?" If we wish to seek thoughtful public judgments about priorities, or how to allocate scarce resources among various kinds of technology (e.g., research on solar energy, applications of Recombinant DNA, or nuclear waste disposal) or how to divide the pie between science and technology expenditures, we must apply more subtle and powerful methods. Using conventional opinion poll methods to ask, "Would you prefer A, B,

---

* Those present at the seminar included William A. Blanpied, Lawrence Bogorad, Harvey Brooks, Peter Buck, Robert S. Cohen, James A. Davis, John Doble, Loren Graham, Stephen Graubard, Gerald Holton, Thomas Kuhn, Marcel La Follette, Leo Marx, Bruce Mazlish, Alexander Morin, Kenneth Prewitt, Derek de Solla Price, Walter Rosenblith, Barbara Rosenkrantz, Willis Shapley, Leonard Simon, Eugene Skolnikoff, Raymond Vernon, and Christopher Wright.

or C" is not a sound or feasible method for eliciting quality indicators that reflect thoughtful public judgment.

The issue of public involvement is usually defined in the form of two questions, each reflecting a different point of view. One is, "Should the public be involved at all in this type of decision?" The second is, "If the public is involved, what form should the involvement take?" One form of involvement that has been recently mandated into law entails public participation in hearings. Another form is the use of public opinion polls to get at people's general views. A third form would be that proposed by Kenneth Prewitt, that is, to engage the interest of only the public "attentive" to science.

In discussing the results of a recent NORC (National Opinion Research Center) survey on public attitudes toward science, commissioned by the National Science Foundation,* Prewitt argues that the scientific community should seek the active support of what he calls the "attentive public," a somewhat elite group of citizens who are fairly well-educated, interested and relatively well-informed in science. Furthermore, the scientific community should relate to this group as a constituency. The purpose here is to provide a source of support for science in some of the controversies that will arise in the near future.

Prewitt believes that such controversy is inevitable—with respect to allocation of resources, the predominance of technology over science, the role of the scientist, science and technology policy, and the general rise and intensity of single-issue politics related to technology and science. I completely agree with this assessment. The 1980s are likely to be a period of polarization and intense conflict of a kind we have not seen in the United States for some time. Furthermore, it is likely that science and technology policy will be involved in those controversies. Prewitt's expression of concern with this issue, and with finding a sound way to deal with it, makes sense.

Prewitt goes on to argue, however, that the proponents of points of view destructive to science come largely from the mass public rather than from the attentive public, and that science needs (or will need) the backing and support of a powerful constituency such as the attentive public provides. This is a plausible conclusion, but I

---

* See *National Survey of the Attitudes of the U.S. Public Toward Science and Technology*, Volume 1: Final Report (National Science Foundation, May 1980).

suspect it is unsound, both as a political strategy and from the point of view of indicators of the quality of science. I believe the mass public *should* be involved, that it has a stake in these issues, and that in a democracy, you must involve the entire electorate in matters of such vital importance. I would also raise questions about the forms of involvement that Prewitt suggests. Setting forth the two contrasting points of view—Prewitt's and my own—may provide a useful framework within which to discuss a new research project designed to uncover the criteria that average Americans use to assess the impact of science and technology on their lives.

The difference between Prewitt's perspective and my own is not on matters of technical expertise (how best to conduct surveys or what the survey findings were). I totally agree with Prewitt on the substance of his research findings. They correspond closely to results of similar research we have conducted at Yankelovich, Skelly, and White. But I do differ with him on the inferences that can be drawn from these findings.

Prewitt's findings show that the public holds science and scientists in very high regard: by a ratio of 7 to 1 the public feels that the benefits of science and technology outweigh the harms, and contribute to the national prestige (which is even more important in today's political climate than it was when the Prewitt study was conducted). His research also shows great public confidence in science's ability to meet certain practical goals, such as predicting earthquakes or getting cheaper energy. It also shows no widespread desire for "populist control" or "participatory democracy." Prewitt sums up: "there is no serious erosion in the confidence of the American public in science and technology," a conclusion at variance with the conventional wisdom that prevailed at the time the study was done. He also quotes a Yankelovich, Skelly, and White study showing that 80% of the public believe that science and technology bring more benefits than problems—a finding confirmed in subsequent research.

Prewitt's findings also suggest that the extent of attentiveness, concern, and involvement of various sectors of the public varies so considerably that it is possible to distinguish (as he does) between an "attentive" and an "inattentive" public. The "attentive public" is defined by three criteria: (1) high levels of interest in science, (2) high levels of current information about science, and (3) commitment to a certain pattern of information acquisition. The attentive public, as defined by these characteristics, is highly correlated with education, to a lesser extent with gender (more men than women), and hardly at all with age. Also, the attentive public has grown considerably, from 8% in 1957, to 18% at the time the study was done. It is consistently more favorable in its general attitude toward science, in the sense of having a less restrictive attitude.

Prewitt then states that we must avoid the "crude utilitarianism" inherent in political democracy, that is, the view that the main criterion for judging the quality of science and technology should be quick payoff. He sees this view as inimical to the values and interests of the scientific community. As I understand his argument, it is that political culture in a democracy is necessarily based on the concept of "deliverables," that is, "quick payoffs." A democracy, he argues, develops practical, utilitarian perspectives so that as a scientific culture becomes contaminated with political culture, a "quick-and-dirty payoff" form of utilitarianism is stressed. As science and technology become more intrusive and more dangerous, this attitude is likely to grow....

...Now, I'd like to state my own contrasting position. Recent studies of the changing attitudes and values of the public lead me to a somewhat different pattern of inferences. For example, we have now in the U.S. (and have had for some time) a growing pattern of skepticism about our institutions, including skepticism about the "wonders" of science and technology. Let me cite just one or two facts from some data on changing attitudes toward quality-of-life issues as they are related to growth.* On the question of confidence in technology (the idea that technology will find a way to solve the problems of society) a bare majority of the public (52%) has that kind of confidence. The older and less well-educated a person is, the more likely he is to have that confidence; and the younger and better-educated, the *less* likely. The range of differences is quite striking. For example, among people older than fifty, this faith is held by 62%; among those who are now in college, it is held by 29%.

In the last decade or so, various polls have explored aspects of the public's scientific world view. In 1970, 30% of the public believed that "everything has a logical scientific explanation"; another 42% said they used to believe that all the mysteries of life would eventually

---

* For a fuller discussion of this point, see Daniel Yankelovich and Bernard Lefkowitz, "The Public Debate on Growth: Preparing for Resolution," *Technological Forecasting and Social Change* 17 (1980): 95–140.

be explained by science but now believed that some things could only be understood in a non-rational way; 28% said they believed that life as we know it is controlled by "strange and mysterious forces that decide our fate." It is interesting to note the direction of the changes that have occurred in these perspectives within the last decade. There has been a modest reduction in the size of the group believing that everything has a logical scientific explanation (30% to 27%), while the group of people saying they used to believe that the mysteries of life would eventually be explained by science, but no longer do, grew from 42% to 48%. The better-educated people who are immersed in the technological outlook are the ones most likely to move toward embracing nonrational explanations, and to question some aspects of the "scientific" world view. These data show a measurable, significant shift that many other research findings support, a shift in Americans' cosmological outlook, an erosion in the consensus view of unqualified belief in science and technology as an instrument of growth and progress.

The causes relate to a profound shift in the structure of Americans' shared values and social norms. All of us are aware of the enormous shifts that have taken place in the domain of marriage, divorce, and family life. The magnitude of the shifts is simply astonishing. In the late 1950s, for example, the University of Michigan found that 80% of Americans believed that a woman who did not marry must be sick or neurotic, that there must be something wrong with her. That belief is now held by a mere 25% of the public; the other 75% accept not marrying as perfectly normal.

The change in outlook with regard to science, technology, progress, and growth is not as dramatic, but it is nonetheless significant. For example, among college students, the level of belief in the idea that "hard work always pays off" has slipped from 73% in the late 1950s and early 1960s to about 43% in the 1970s, from majority to minority status. Most college students are still confident that, in exchange for hard work, they could gain the conventional payoff (that is, married life, a home, an automobile, and respectability) but they do not value it as much as students had in earlier periods and do not believe that it is worth the effort.

My hypothesis is that in the 1980s we will undergo intense social stress because of cultural changes born in the 1960s and 1970s. These changes have now provoked a fierce reaction in movements such as the "Moral Majority" and among people who feel that the fundamental premises of their lives, having to do with family and church and religion, have been challenged. It is one thing, however, to have such a battle rage in a politically and economically stable society, and another to have such changes superimposed on a society experiencing economic stress due to inflation, due to decline in the median income, and due to the fact that inflation is redistributing income in forms that make people feel there is no social justice. In the 1980s, this cultural backlash will be superimposed on an exacerbated conflict between the "haves" and the "have-nots" and also superimposed on conflicts between young people who work for a living and the growing number of older people who are living on social security. Such conflicts tear at the fabric of our society.

Because these changes have occurred so rapidly, they have not occurred uniformly. At various levels of American society, one can find traditional values in almost pure form, new radical views on quality of life at the other extreme, and a vast in-between. Some of these new views are expressed with theological intensity because ultimate values are at issue. When the Moral Majority discusses evolution, for example, the issue is not so much a controversy over the factual bases of creation as an emotionally charged symbol for a whole set of moral and religious values.*

Science and technology will not be permitted to stand aloof from this values controversy. On the contrary, they will be plunged into the middle of it. My concern, coming back to Prewitt's conclusion, would be that if the scientific community tried to make common cause only with the so-called attentive public, it would be enlisting on one side of a social-class battle, because the "attentive public" is not so much a scientifically minded cross section of the public as it is an educated elite who share certain outlooks and values with the scientific elite. To make common cause with them readily implies that the mass of the public, the 82%, are presumed to be the enemy. I think this is bad politics, because it politicizes and contaminates science and technology issues with ideological, social-class issues. It would so confuse the situation that science and technology would be put on the defensive, because the majority of the public—a riled-up majority—would then be put in the position of becoming adversaries.

* These and other on-going changes in American values are discussed at greater length in Daniel Yankelovich, *New Rules: Searching for Self-Fulfillment in a World Turned Upside Down* (New York: Random House, 1981).

It also presupposes that the evaluative criteria of the 18% "attentive public" will be less crassly utilitarian than the criteria of the less knowledgeable 82%, which may be an incorrect assumption.

My quarrel with this approach is that, for narrow scientific issues, the "attentive public" doesn't know enough science to serve as a jury of peers which can make decisions on a sound technical basis. To the extent that these are political values, why exclude the mass of voters? The split, in other words, between the attentive public and the mass public is based on a supposition of shared values—a premise that is not a sound basis for the issues we are looking at. Why not involve the entire public? Why not bring into the picture those people who do not go to science museums and do not read science magazines and do not do some of the things that the attentive public is supposed to do, but whose lives and futures are just as affected by these technological decisions as the people in the so-called attentive public? . . .

Everybody who has studied this issue has found that it is extremely difficult to get the public to distinguish between science and technology. It is almost impossible to do it with a conventional public opinion poll. It may be possible to do it with a technique I am currently testing. We bring people together as a group and give them a chance to gain some background and understanding on matters of policy with respect to technology.

A policy with respect to the uses of technology, for example, often concerns the allocation of resources. Presuming that our society's resources are scarce and that we cannot do all the things we want, to what projects shall we give priority? Matters of choice on issues like health and energy turn out to be value choices; these are issues that relate to the kind of values that people want the society to represent, the kind of society they want to live in. Those values are often hidden or are defined in technological terms when they should be defined in value terms. In the debate over nuclear power, for example, there is a deep interpenetration of technical questions about safety with value questions about what kind of society should exist and how technology should be used to achieve it. Those are fundamentally political issues, which require public participation. Our project will try to find out what criteria people bring to bear when they make judgments about where to allocate limited resources.

The research method involves two instruments. One is a Personal Priorities Inventory for making judgments about science and technology; this is a quality-of-life scale which asks people to state what is truly important to them personally. There are several subdivisions. One pits the quality-of-life values against utilitarian instrumental values, the "small is beautiful" kind of concern. . . . There is also a set of risk-benefit trade-offs. How much risk and danger will people accept for the sake of economic growth, or national prestige? There is a set of more irrational values, for example, "freedom from Ralph Naderism" (that is, the freedom not to have people tell you what's good for you). There is a very strong strand of anarchic feeling in the society today. Studies show an intense concern over the theme of loss of control. People feel that they're losing control over when or whether they will retire, maintaining a stable income against inflation, over their lives in general. These fears certainly affect their judgments on value choices. Our inventory of questions attempts to get people to position themselves on the values most important to them at that moment.

The second instrument includes six or more specific proposals for research on science or technology. These proposals incorporate different kinds of payoffs—industrial payoffs, consumer payoffs, power for the society, knowledge results, human results, beneficent results for all humankind. They entail different levels of risk and rewards. They make different promises within different time scales. They promise benefits, some of which are murky versus some that are very clear-cut. They deal with the controversial and the non-controversial. These proposals also represent issues that members of the science and technology community feel should be pursued or should get a portion of available resources. The proposals have also been reviewed by specialists and by representatives of the various sides discussed. These two instruments—(a) the values inventory, and (b) the specific research proposals and the arguments for and against them—are written in terms that average people can understand . . . .

In sessions that last for several hours, we review each of the proposals and give each group the chance to argue and discuss as if they had to make the decision themselves. Each group of fifteen acts like a jury, using social interaction to arrive at "public truth." . . .

The important thing is to give people a chance to argue, to interact, to make explicit the basis on which they prefer or reject each proposal. The socialization process is such that, when the pros and cons of each proposal are made explicit, the discussion moves from technical considerations to value considerations. The emphasis is on judgments related to the kind of life we want to live, arguing the pros and cons, making explicit the benefits and the risks in human terms. We are not consulting the public to get a technical judgment, but to get a political judgment.

The goal of the project is to have people "work through," argue, fight about those proposals, and come to some consensus at the end (if possible). We can then measure the changes that have taken place between the initial and later votes, and gain some understanding of what moved people to change their minds. Using that design, a number of interesting results can be obtained. First, we gain a picture of how people respond before they have had a chance to engage the arguments and hear other people's points of view. Then we gain a picture of how much they have been changed by the arguments. It really puts the values to the test and makes them explicit. We get the *considered* judgment of the public rather than a top-of-the-head, top-of-the-mind judgment....

Finally, let me stress the importance I give to finding some way to replicate the social process of decision-making. If the public is going to be involved in policy decisions, people must have a chance to wrestle with the proposals, to have a true dialogue with their peers, rather than simply have their superficial judgments recorded. This process of dialogue and argument will be better for science and technology. The results will be less crassly utilitarian than the approach that attempts to persuade the "thoughtful opinion leader elite" to support particular proposals in political battles. Admittedly, this approach expresses a faith in democracy and a faith in the concept of an informed public. In the context I have described, "informed" does not mean the attentive public, does not mean only the people who are knowledgeable about science and keep up with it. It does mean "informed" in the sense of having the opportunity, within a structured setting, to face alternatives and to make explicit the value premises in terms of what those alternatives should be. From the point of view of making quality indicators explicit and of giving science a firmer, larger political base on which to base future support, that is certainly better than an approach on a "class warfare" type base....

Editor's Note: In the discussion following Mr. Yankelovich's talk, the participants focused on some of the difficulties involved in measuring public attitudes to technical issues that are undergoing rapid change. Excerpts from the dialogue directly pertaining to this topic are included below.

Some social scientists are attempting to correlate media coverage of controversial science or technology issues with data gathered in public attitude surveys. One researcher, for example, has shown that when media coverage of a controversy (and, hence, the amount of information available to the public) goes up, then public opposition to a proposal also rises; when media coverage decreases, so does the recorded public opposition.* In the discussion period, I asked Mr. Yankelovich whether or not the fact that people are presented in these groups with more (often new) information on the various research topics could bias the panels toward opposition *of the research.*

Yankelovich Certainly when people get more information, and think through issues, opposition can firm up. ... Discussion can polarize as well as create consensus. One interesting example involves public attitudes on the Panama Canal treaty. When the Carter Administration found that most of the people who were opposed to the Panama treaty were opposed because they lacked certain information, they mobilized an information program on Panama; it had two results. One was what the Administration hoped would happen: as people got more information, the number of people in favor of the treaty increased. But something else happened—the opposition also solidified. Both things happened at the same time. That is one of the effects of thinking something through. What you get is the distillate, the precipitate of people's thought-through views, and they are different from the views people have before they start the working-through process. This effect could work in both directions.

Alex Morin made a very important point earlier in the discussion regarding people's tremendous need to be consulted. The act of bringing them into the picture and consulting them has an effect by itself. Part of that effect, within a group, is for the group to want to seek accommodation rather than obstruction. Once people are consulted, there is an impetus in the dynamics of the group process to see if some sort of common ground can be found. If there is any bias in the group dynamics, it will be a bias toward consensus—which is why I think it will be interesting to contrast viewpoints before people are given any information (because that approximates the state of the public in general) and their views after participating in

---

* See the remarks by Allan Mazur in "Mass Media News Coverage of Scientific and Technological Controversy," 6 *Science, Technology, & Human Values* 36 (Summer 1981): 29.

that group process, to look at the nature of the change and the dynamics that underlie it. From the research findings you would learn two things: one, the pre-workthrough response, and the other, the post-workthrough response.

Stephen Graubard (Editor of *Daedalus*) In your remarks on democracy, you are entering a field that is mined with elements of controversy. There is a substantial question of not only whether there has ever been successful public pressure of the kind you describe, but also whether democracy at any time has been capable of the kind of responses that you indicate. Are you starting with a notion of ultimately affecting opinion (and even action) in a way that for a hundred years or more theorists preoccupied with democracy have thought to be impossible? . . .

Daniel Yankelovich That is a difficult question. If my assumption about the 1980s being a troubled period is false, then the results of the exercise may have some theoretical interest but possibly not much practical value. . . . On the other hand, if the 1980s do turn out to be a societal cockfight, with science and technology in the middle, and if the political pressures loom urgently, a whole range of questions will arise about how best to respond, to defend, and about what techniques of involvement to use. Almost inevitably, political pressures through Congress or through an articulate group like the Moral Majority or some other self-appointed interest group will tend to dominate . . . .

If there is turmoil, then we may have opened an avenue of insight into the public mind and how it engages these issues which will be of keen interest—and great surprise—to the scientific community. The result may be discouraging in some respects, encouraging in others, but certainly different from the stereotypes and expectations.

There are profound advantages and dangers in democracy, but they all reduce to one rather simple point. In a free society, where citizens have the ultimate power, if you fail to consult or involve them, and if they can vote, then you can provoke all kinds of mischief and trouble. In our kind of democracy there must be some viable method for making people feel like responsible citizens. Should we not have some insight into how the public does think? What would happen if there were this kind of consultative process on issues that involve the public most deeply? . . .

... The research strategy is dictated by data indicating that people do not feel qualified to make judgments, and that they believe certain questions are for the experts to decide. There is a tendency for an individual to divide issues into "those appropriate for the experts because they are technical" and "those appropriate for me because they concern the way I want to live my life and the kind of society we are going to have." ...

[Responding to a question, Yankelovich continued:] You suggest that if issues occur in the real world with five sides, we should try to see the extent to which people can wrestle with them. That strategy, at least, leads to a useful insight. In ascertaining information and getting the public involved, there is a certain cut-off in terms of level of complexity, but we may wish to see what level of complexity the public can engage rather than assume in advance that the complex cases cannot be gauged.

... I think you can force people in these groups to make decisions by saying "We're talking about an issue where the range of controversy in the scientific community is so great that for almost every scientist involved there is a different opinion, and they range from A to Z. Nevertheless, even though the scientists disagree, we want you to come to grips with this issue and make decisions about it, because we want to see the basis on which you make them." Now, if under those conditions, you cannot force people to come to conclusions, that is an interesting finding. If you can, and you see what the basis is, that is also interesting. ... There is a large group of citizens who say about these matters, "Why are you talking to me? Why are you consulting me, what does it have to do with me?" You cannot get people to engage with a problem until you answer that question.

... In the 1960s and the 1970s, the majority of the public felt they could have more of everything and, furthermore, that they were entitled to more of everything. Now people fear they may end up with nothing. These two states of mind co-exist even though they are contradictory. People are increasingly beginning to realize that they cannot have more of everything, *and* that they may not be entitled to more of everything. There is beginning to be suppression, if not repression, of certain desires. In the 1980s we will, I suspect, see the evolution of a new social ethic, one that is neither the old self-denial Protestant Ethic nor the "duty to self" ethic that momentarily replaced it in the 1960s and 1970s.

Any new social ethic, in our present condition, is bound to be unstable. The situation produces a moving target. The project I have described attempts to look at these values as they change in relationship to emerging technical issues, with the understanding that the process is going to be dynamic and modified but that, at any one moment, we can gain a good understanding of where the public stands in relationship to these values, and how it employs them in assessing specific proposals for uses of technology. . . . You can think of it, as I do, as a cultural revolution or as mere drift, but the culture in this society *is* changing. Within that framework of change, the issues of science and technology are central. There is a change going on in Americans' world view, and science and technology are at the center of that change.

The Social Assessment of Science, or the De-Institutionalization of the Scientific Profession

Peter Weingart

Public criticism of science and technology may not be new, but its character is changing in breadth, intensity, and focus, one example being that the target of public criticism has moved from the application of science to the research itself. Historians might point to instances when that has happened before, but at present it seems as though we are facing a new phenomenon.

Let me classify three different manifestations of this public assessment of science: public challenges to research autonomy; government intervention through regulation and control of research; and professional self-control, which takes on new characteristics under public pressure. Although these types of assessment are, of course, interrelated (one may be the cause of the other), we can also discern two significant trends. First, the mechanisms of professional self-regulation are increasingly supplemented by external (that is, social or political) regulatory procedures—a development that implies, of course, partial loss of autonomy and control in all the professions but particularly in science. Second, traditional political institutions are being severely challenged (perhaps more in Europe than in the U.S.) as formal mechanisms of decision-making are supplemented by informal processes of regulation and control which derive their legitimacy from local protest groups, from *ad hoc* movements, and, most important, from demands for public participation, demands which are linked to specific, often short-lived issues.

These trends indicate the framework for the questions I want to address here briefly: What intellectual and social factors account for the sudden eruption of demands for assessment of science? And what implications do these trends hold for science as a profession?

The history of the institutionalization of science is, in a way, a history of the extension of the social space of the pursuit of knowledge for its own sake. At its beginning, the demarcation drawn by the seventeenth-century academies between science and the spheres of religion and politics is a lesson for that hypothesis. The negotiation between religion and science (or, more generally, between society and science) led to different results depending on cultural and political conditions, but, as a general pattern, the process of negotiation has continued throughout the history of Western science. The modern German constitution, for example, includes a section on "freedom of research" as one of its basic paragraphs, something apparently unique in the world. Research had not fared well under the Nazis and this section was written as a reaction against that repression. The result is that "freedom of research" exists now as written law. Historical experiences, therefore, *do* have an effect on this type of negotiation process. But this negotiation process often brings science into conflict with other social institutions—why?

Max Weber, looking upon the victory of rationality in the world, alarmed his readers to the fact that, although rationality may be victorious, the result was the erection of a new "Iron Cage." With the success of science, the expectation of gaining truth, understanding, and salvation (and, of course, what the German word *Sinn* conveys) had been disappointed. The sciences provided direct guidance for human action; but, in providing instrumental rationality, they also helped to dissipate the guiding images of other institutions. Let me give a few examples. In the law, the influences of science (or, in particular, sociological and psychological evidence) have, in fact, helped to dissolve the link between crime and punishment (which is a very deeply rooted association). Crime is now explained in terms of individual biography, as a problem of deviant behavior, and deviant behavior is explained by conditions of, for example, childhood socialization and is no longer subject to moral judgment. Social integration, as the institutional mirror, has become a scientific, not a moral, problem. Institutionally, decisions of punishment are therefore often referred to expert wisdom or expert advice. A similar example may be found in the field of medicine, where the borderline between the normal and the pathological with respect to mental illness is determined by psychology, psychiatry, and psychoanalysis—sciences in the wider sense of the word.

The institution of education provides perhaps the most dramatic example: the construction of educational theory and curricula that aim at reproducing or simulating practically all areas of life, or at supplementing experience. In other words, it is no longer experience that selects criteria of importance for what we do; instead, scientifically constructed curricula are used to prepare children (and even adults) for such normal human tasks as mating, health care, feeding, child-rearing, aging, etc.

In cases such as these, the problems are increasingly removed from the orienting functions of traditional institutions and defined in terms of science—a process I would like to call "scientification," despite definitions of scientification used elsewhere. This scientification, in replacing the traditional orienting functions of institutions, has its correlate in what could be called "de-institutionalization." In other words, traditional institutions are being dissolved by popular preference for scientific knowledge. As knowledge pertinent to the problems governed by institutions is being diffused, the institutions lose their orienting power. This process has also been enhanced (at least for the last 50 or 60 years) by the general reliance on science's explanatory potential. Governments, for example, may know that some of the problems of implementing a technology have not yet been solved, but they count on the fact that science *will* solve those problems (for example, the storage of radioactive wastes from nuclear power plants) when the technology is implemented.

Science, of course, cannot orient human action. Although science undermines the orienting function of other social institutions, it cannot replace them. For instance, it reduces problems of fact to the underlying problems of value, but it nevertheless remains ambiguous with respect to those values. To give an example, determining the carcinogenicity of certain industrial emissions may clarify an issue, but it cannot decide it because certain risks and benefits are involved in the decision to reduce emissions, and these risk and benefit assessments have to be made elsewhere. Thus, the scientification of social institutions is bound to a politicization of science, which is an inevitable outcome of the extension of the boundaries of science.

This politicization has very serious consequences for science as a profession. Let me briefly remind you of the definition of a profession. First, there must be a specialized body of knowledge, either scientific or systematic; second, the profession must provide training, must socialize its students; third, there must be a procedure for granting access to practice, through exams, titles, etc.; and fourth, there must be an institutionalized code of conduct. A profession defines the border between experts and laymen, and it gives special services to a specified clientele. Science, however, is a unique case. As has been pointed out repeatedly, science does not have a specific clientele; and, traditionally, science is not under the control of a clientele. In other words, no clientele determines the relevance of what it does; rather, this is done within science by science itself. Insofar as this situation is true institutionally, then the autonomy of science exists; insofar as this is not true, it no longer exists.

Challenges to quality or relevance are usually considered by an institution to be illegitimate interventions into its autonomy. The question is why, and under what conditions, does a profession (and science in particular) lose the privilege of this quasi-autonomy? First, when the knowledge being produced comes into conflict with the values of the clientele, or with the values of society at large. For example, one type of question posed recently for the biological, biochemical, medical, and behavioral sciences concerns the definition of appropriate research subjects. I would assert that in such cases what is taking place is the assessment of research itself, not merely the assessment of a specific technology or of the applications of science. It is in part because of this extension of science into the most sensitive realm of ethical convictions that the assessment movement has developed. This extension may also explain the viciousness of the conflicts that do take place.

The second reason that a profession may lose its privileges of autonomy, it seems to me, is partly a result of the first—that is, an internal consensus is lost. When research touches upon values, the moral and political convictions of researchers are activated. Conflict is no longer confined to disputes between experts and laymen, but occurs also inside the scientific community, within the borders or the institutional limits of the profession. The dissenters then may appeal to the public, through warnings of hazard, for support of their critique—appealing, in other words (and Culliton has, I think, defined this very well*), to the public for political views held within the profession. The consequences of such actions are obvious. The profession loses authority over the production, application, and interpretation of knowledge. And with the politicization of expertise, the expert function may also be lost.

The third general reason for loss of autonomy is that quite definite

* See Barbara Culliton, "Science's Restive Public," *Daedalus* (Spring 1978): 147–156.

changes are taking place in the social and political order. The expansion of government regulatory functions, for example, is a process independent of the development of the professions; it has to do with the increasing complexities of governmental administrations. This process implies an increasing demand for knowledge used for legitimization, aside from its instrumental value. Government thereby becomes the chief client of science.

So, for science, the situation has changed from both the inside *and* the outside. The pressures to produce knowledge for instrumental and for legitimizing purposes no longer allow scientists to retain professional autonomy, distance, etc.

The other trend affecting the public assessment of science is the general decline in the legitimacy of political institutions and, with it, the emergence of participatory and protest movements, often in connection with science and technology related issues. The familiar, but seemingly paradoxical pattern is that the instrumental quality of scientific knowledge makes it a tool for legitimating policies, but then its ambivalence with respect to value decisions renders it useless. Opposing groups take recourse to scientific experts, who give contradicting advice because of the political nature of the issues involved. The consequent loss of professional authority makes the profession vulnerable to criticism and challenge, which destroys its very function and, as a result, the state takes over in regulating the profession. This process has been very visible in biomedical research, for example, as in the dramatic example of Recombinant DNA research.

In conclusion, we may ask if the assessment movement is destroying science as a profession, as many fear it will? If scientific and technical expertise are replaced by lay knowledge, will science *for* the people then become also science *by* the people? And, above all, could that work?

The pattern that I've tried to describe is one of science increasingly politicized, with conflicting factions and publics. Although its legitimizing power is decreasing, it is still being used as a resource in political debate and in dealing with social problems. The public assessment of science may, thus, be interpreted as that stage in social development where the orienting quality of science as an institution has been exhausted, the ambivalence of its instrumental rationality with respect to values has been uncovered, yet at the same time, its usefulness in solving problems, and for understanding the complexities of the world, cannot be discarded. Science *for* the people, therefore, may in a loose sense be realistic; science *by* the people is not. The assessment movement may mark the end of science as an autonomous profession and the end of scientism, but not the end of science as a specialized social activity.

Note

This paper is an edited version of a talk given by Professor Weingart in the STS Colloquium series of the Program on Science, Technology, and Society, MIT. An elaborated version will appear in German under the title "Verwissenschaftlichung der Gesellschaft—Politisierung der Wissenschaft," in W. van den Daele, W. Krohn, and P. Weingart, eds., *Legitimationskrise der Wissenschaft* (Frankfurt/M).

The Quality of Science Equation

Orrin G. Hatch

It may be a fact of life that there is good and bad in everything. It is perhaps the true test of all men and women to make proper assessments of what is good and what is bad. With science it is no different. The quality of science is held up by the sturdy triangular foundation of individual scientists, the research work itself, and national policy toward science. Each of these components should be examined separately because they are each dependent variables in the quality equation.

The equation will never balance without good individuals. While individual scientists possess varying degrees of ability, creativity, and insight (which alone could be the topic of a paper much longer than this one), there is another factor which he or she should contribute to science—integrity. Even the very best science should not be conducted by Dr. Frankenstein. Society may gain significant scientific knowledge from such research, but if this knowledge is not gained truthfully and with good will, then, to borrow from Paul's first letter to the Corinthians, "it profiteth nothing and will fade away." Our society as a whole may have become more prone to put integrity last, below intelligence, below perseverance, and below congeniality, but it is a virtue we ought to put first.

Appreciation for right or wrong ought to begin in the home and should never be compromised or confused throughout a person's life. Our colleges and universities have a monumental task in embuing those who enter with the idea that their education does not entitle them to special privileges as much as it places upon them special responsibility for the upkeep of what is good and the uprooting of what is bad in the world. My own alma mater, Brigham Young University, has inscribed on one of its gates: "Enter to Learn, Go Forth to Serve." Our nation and our world depend on scientists who will "go forth to serve." While every person in every profession and station in life should be reawakened to both contribute and demand these qualities in others, scientists may be singled out since they hold the key to the future. Let us at least begin the assessment of good science with the assurance that there are good scientists dedicated to thorough, objective, and honest research.

Good scientists are important for the same reason that good auto mechanics, good doctors, or good building contractors are so important. People in these occupations could take unfair advantage of those who know less. An auto mechanic could recommend expensive brake work; a physician could suggest that surgery is necessary when it is really elective. Congress relies on the expert advice of many good scientists to explain the difference between software and wash and wear. Congress simply cannot deal in depth with the vast number of issues in health, education, economics, defense, energy, environment, foreign affairs, commerce, welfare, and government organization. Congress does not have the luxury of time to study each separate proposal in science and make an expert determination that a project for laser research represents better science than one looking into thermonuclear fusion. For this function, the nation depends on peer review.

There can be no denying that a wealth of ideas are waiting to be tapped and converted into experiments and knowledge. A wise man once said that "you have to be pretty smart to know how much you don't know." As we approach the year 2000, we can both appreciate how far we have come and how far we must go. Many scientific roads have not yet been travelled. The future depends on the identification of the most promising roads to discovery and the excellence of the methodology to achieve new information. This is good science.

The role of Congress is to develop good policy in science and technology which will best serve the overall national interest. It is within the reasonable purview of Congress to set disciplinary priorities, decide what agencies of government should have administrative responsibilities and allocate an appropriate level of funds for Federal research activities. The major questions of policy include the optimal combination of basic and applied research and innovation, science education, the export of high technologies, excessive regulation, stimulating private sector R&D, and improving the scientific instrumentation and facilities available at our country's universities and colleges. Congressional attention is also directed toward the marginal costs

The Quality of Science Equation 121

and benefits of continuing the space program and a crash Federal program of research in energy.

A good policy, of course, must not only advance our country's position in science, but also complement the broad policy objectives of the United States. Congress makes such policy choices on behalf of the American people. The choices concerning science policy do not necessarily reflect Congressional assessment of what is good or bad science. The fact that less money will be available in FY1982 for the social sciences, for example, should not be interpreted to mean that Congress perceives all social science research as bad science. Nor should increases in the mathematical and physical sciences be hailed by those disciplines as proof that they propose only good science. A new era of budget restraint will not permit across-the-board real growth as has long been the case, and it is critical that members of the scientific community be able to adapt to shifts in policies and priorities without offense and without focusing undue attention away from the research underway. Congress intends to cast no judgment on what is good science or bad science, although good scientists should not mind explaining some of the less obviously efficacious research expenditures. Good scientists will also understand if Congress determines some restriction on Federal research dollars or some procedural change is needed to insure the integrity of the taxpayers' investment in good science.

The United States must continue its active pursuit of scientific knowledge and technological development. There is no doubt that Federally supported programs through the National Science Foundation, National Institutes of Health, National Aeronautics and Space Administration, and the Departments of Defense and Energy have contributed enormously to our efforts in this regard. These instruments of Federal policy help make good science possible. A good policy leading to American eminence in science and technology may, however, like once fresh food, become old and unusable and should be reevaluated periodically to meet new goals. Congress always seeks the best policy for science and technology—one capable of inspiring the best scientists to perform the best science.

The Role of the Federal Government in Supporting Research and Development

George E. Brown, Jr.

A major science policy decision is pending in Washington, DC, which is obscured by debates over Federal budget priorities and deficits. The decision may very well be made by accident unless a wider audience becomes aware of the issues involved and, in turn, becomes an active participant in the debate. Put starkly, the question facing the country is, "Should the Federal government phase out support for non-defense science and technology?"

Before I continue, let me make a few disclaimers. I am a Democrat who believes government can be part of the solution to the needs of society. The Reagan Administration is Republican, and dedicated to fundamentally altering this historical role of the Federal government. To date, Administration spokesmen have professed support for Federal research and development efforts, while making the largest cuts in the history of our nation. These facts color what I am about to say, and are important for the reader to know. However, there is no substitute for independent verification and evaluation of the facts.

The reason I phrase the question, "Should the Federal government phase out support for non-defense science and technology?" is that this is the trend and the logic of the Reagan Administration's philosophical and long-range budget posture. When one looks at long-range budget projections, the growth in the defense budget and the reduction in the discretionary (non-entitlement) parts of the civilian budget, pressure on research and development support grows greater every year. The FY1982 cuts which President Reagan proposed in September 1981, only a few days before the budget was to go into effect, hit research and development hard, and came on top of cuts proposed in March 1981. These two rounds of budget cuts came in the first nine months of the Reagan Administration. The trends for the future look far bleaker as the pressures for additional cuts grow, and as the relatively defenseless civilian research and development programs look less essential.

Many of those who are making policy in the current Administration are proponents of the economic philosophy of Nobel laureate Milton Friedman, who argues that "the National Science Foundation, the National Endowment for the Humanities, and tax subsidies to higher education are all undesirable and should be terminated." Even more members of the Reagan Administration expound the social and economic philosophy of George Gilder, whose book, *Wealth and Poverty* (New York: Basic Books, 1981), is considered mandatory reading by President Reagan's political appointees. In that book, George Gilder makes the alarming assertion that "most scientific breakthroughs are made by men in their twenties or early thirties. The national laboratories, the Food and Drug Administration (FDA), the Environmental Protection Agency (EPA)—all used by government to appraise the products of civilian science—are full of men who are past their prime emotionally and intellectually committed to earlier technologies, and deeply resistant to progress."

We have come a long way since Vannevar Bush's 1945 report to the President, *Science—The Endless Frontier*. It is clear that the entry of the Federal government into non-military and non-agricultural research and development support in the post-World War II era changed this country.

Bush's report, delivered to President Truman, is still eloquent reading today. In fact, it is surprising how many of the problems grappled with in 1945—such as excessive government secrecy—remain important issues today. Let me quote one central passage from the Bush report which concerns the topic of my commentary:

Science Is a Proper Concern of Government

It has been a basic United States policy that government should foster the opening of new frontiers. It opened the seas to clipper ships and furnished land for pioneers. Although these frontiers have more or less disappeared, the frontier of science remains. It is in keeping with the American tradition—one which has made the United States great—that new frontiers shall be made accessible for development by all American citizens.

Moreover, since health, well-being, and security are proper concerns of Government, scientific progress is, and must be, of vital interest to Government. Without scientific progress the national health would deteriorate; without scientific progress we could not hope for improvement in our standard of living or for an increased number of jobs for our citizens; and without scientific progress we could not have maintained our liberties against tyranny.

The Reagan Administration may do this nation a great favor by its ideological excesses in cutting the budget. As we all know, it takes a crisis to consider radical changes. Inefficiencies do develop with complacency. On the other hand, unless those who understand the need for a strong role by government in the civilian scientific and technological enterprise speak up and make their views effectively known to the Reagan Administration and the Congress, the "irreversible damage," feared by Frank Press and others who attended a recent scientific summit meeting at the National Academy of Sciences (NAS),* is likely to occur.

Ironically, I find myself somewhat in disagreement with the remedies of Press and others who attended that NAS summit meeting. In their position statement, the participants, who included many research managers from industry, argue, "The proposed reductions in the President's September or Fall budget will do irreversible damage unless longer-term research, in contrast to development and demonstration, is protected." My concern is the unnecessary "contrast" of basic research with development and demonstration. This artificial separation of one component of the research and development continuum (basic research) from the rest (development and demonstration) may easily result in the cannibalizing of one part of the scientific and technological enterprise to feed another.

My own view is that the United States is suffering economically from its schizophrenia over what the proper role of government is in supporting research, development, and demonstration. The U.S. still has the strongest basic research base in the world, and while the policies of the Reagan Administration threaten this base, our economy is threatened today by our nation's competitive disadvantage in supporting new technologies. The Federal programs which were begun under the Carter Administration to help U.S. innovation and productivity were too little, but at least were steps in the right direction. The termination of these programs by the Reagan Administration illustrates a major switch in public policy that has been made without public debate or consent.

*"Federal Conference on Research and Development Budget for 1982 and the Future," held at the National Academy of Sciences, Washington, DC, 26–27 October 1981.

I am not prepared to predict what the Congress, or even the Reagan Administration, will do over the next several years. The votes in the Congress are very fluid, and the Administration has shown an ability to be flexible and alter its position. I am prepared to say that in the absence of broader public debate, and greater interest in what we do in Washington, the Federal government will continue phasing out support for non-defense science and technology. The scientific community should do more than help establish a value system for cutting its own budget.

Decision-Making for Quality Science

Don Fuqua
and
Doug Walgren

In science, as in many other areas, the sharp inflationary pressures and budget restraints of the last few years are forcing a fundamental reexamination of the methods by which investment priorities are chosen. Budgeteers and politicians are seeking to apply cost/benefit thinking to science and technology, recognizing that resources are not sufficient to extend even to all highly credible initiatives.

This type of thinking can, of course, be done retrospectively: the evolutions from solid state physics to microelectronics, and from basic molecular biology to genetic engineering, are classic examples. In each case, high payoff research included the most basic groundwork science. Breakthroughs in esoteric and fundamental areas often proved to be as much or more effective in producing societal benefits than incremental progress in highly developed near-application technology.

Few clues, however, indicate how to predict the benefits of targeted research investment. If key investments are likely to be far from the field of eventual application, and the most beneficial developments to be those which totally transcend conventional technology approaches, predictive capability in a traditional sense is eliminated almost by definition.

No simple formula is likely to be found for managing and directing the resources of science for "quality." Instead, ways must be sought to encourage diversity, adaptability, and rapid cross-fertilization between fields so that emerging ideas may spill over into totally unforeseen developments. At the same time, we must attempt to identify particular areas that have reached maturity and are ready for special investment for immediate social benefit. The historical decisions to invest in a "war on cancer," and subsequent debates on their effectiveness, show how difficult it is to do this wisely.

Although no system or institution is likely to provide a simple mechanism for these choices, society will most likely continue to rely on the scientific community for guidance. Illustrative examples are Einstein's famous war-time letter to Roosevelt on the potential of nuclear power,* as well as many less dramatic instances in the post-war period. Realistically, the balance of skills and sense of up-to-date developments requires that the lay society rely on scientific community consensus when making technical decisions. Science, therefore, is privileged—or perhaps burdened—with an inevitable requirement for extensive self-governance.

The privilege aspect of self-governance may be obvious. The burden, however, may prove to be a high price to pay. The importance and cost of science is so great that the rest of society will not lightly tolerate careless mistakes or self-serving frivolity. Self-governance must be carried out relatively well if some much less effective and more arbitrary system is not to take its place. This means, however, that the scientific community must spend time and energy on relatively frustrating organizational, priority-setting, and consensus-building activities, with resulting loss of time and energy from the more individual and creative discovery and development activities that provide much of the career satisfaction and joy for scientists and engineers. Facing the organizational burdens juxtaposes sharply the contrast between a day in a committee meeting and a day in the laboratory in pursuit of an exciting scientific result.

Society will also inevitably put pressure on the internal scientific decision-making process. This is not new, of course; corporate research directors, military technologists, and regulatory agency scientists have long had to face demands for highly specific and premature cost estimates, development schedules, and numerical risk estimates from their various impatient lay bosses and colleagues. The pressure will grow, however, as the stakes become higher and the competition for funding greater. Perhaps the greatest effort in building the utility of the scientific community's self-governance system should be to build frank understanding on both sides of its imperfection, and a recognition that its chief virtue is only its probable superiority to even more clumsy and nonproductive alternatives. Despite the frustrations, the scientific community must not turn away from allotting at least a reasonable amount of its talent and creativity

---

* Albert Einstein to President Franklin D. Roosevelt, 2 August 1939.

to its self-governing mechanisms and institutions. The alternative imposition of other kinds of governance would be disastrous for both science *and* society.

Equally frustrating will be the necessity for the scientific community to recognize the importance of distinctly non-scientific values, both in the design of its own self-governing system, and in the way society uses its results. Public concern for the welfare and treatment of laboratory animals is an example of the first; public reactions to the ranking of risks such as nuclear power accidents, smoking, and auto seatbelt neglect indicate the problem of the second. In its self-governance, for its own and for society's good, science must do the best job it can in integrating its technical perceptions with other recognized social values. On the other hand, it must not become discouraged or alienated if it finds, as many other groups have found, that its own best perceptions can often be overruled by a broader social consensus based on different priorities.

Pressure for dramatic or simple answers appears to have taken its toll in recent years. Incidents of laboratory fraud, from painting mice to falsifying notebook entries, have been a by-product of the ambition and excitement stimulated by the drive to understand and cure cancer. Broad scientific risk questions, such as fluorocarbon threats to the ozone layer, the possibility of CO_2-induced climate change, and the origins and impact of acid rain, have created a new and largely unflattering image of the scientist. Instead of maintaining their roles as unobtrusive problem solvers, technical people have often been thrust into adversary proceedings. Scientists have, for example, been used as the authority behind slick media advertisements that attempt to persuade the public of the total credibility of one position, and heap ridicule on the opposition. Although the manipulative mass-media approach may be effective in some instances, it is doubtful that the public will accept it as the most comfortable way in which to receive advice from the scientific community.

A key to a more effective decision-making partnership between the public and the scientific community is going to be recognition of the crucial need to up-grade public scientific literacy and public ability to understand and participate in technical decisions. Scientific quality decisions, either in investment choices or in evaluation of risk or opportunity, may best be made largely through various scientific self-governance systems. However, the chance that even the best of these decisions will survive to a full social consensus depends on relatively informed and sophisticated public understanding of at least the broad factors determining them. The ability of the public and the scientific community to integrate each other's perceptions into their own depends on a mutual respect that must be built on some elements of common understanding. The gulf between the "two cultures" may necessarily represent quantitative differences in technical ability, but cannot be as great as totally different qualitative outlooks on the world. As we perceive the impact of astrological advice on the thinking of many individuals, even in the 20th-century world, it is clear that to ignore society's scientific literacy level risks substitutions of the most bizarre approaches for scientific advice in determining the quality of science.

Public technical literacy may be the key to some of the most crucial decisions facing humanity in coming decades. The most vivid example may be that, in countries around the world, the number of people able to grasp the scale and possible impact of modern weapon systems is very small. Yet decisions concerning those systems affect the daily lives and possible existence of billions of individuals. The investments in these systems siphon huge amounts of technical, human, and financial capital from other sectors. Their use on a large scale could render our planet uninhabitable. Tragically, however, most citizens believe technical decisions concerning military strategy to be beyond the scope of their common sense, even in relatively open societies. A screen of unfamiliar technical jargon and concepts prevents them from applying the same pragmatic common sense that has allowed reasonably informed participation in decisions affecting roads, schools, and many other important aspects of their lives.

The decisions concerning military investments and use of their products are now necessarily in the hands of a small technical elite. Many outside observers decry the irony of seeking security through expensive, complex, and often mutually destructive modern weapons which, in a broad context, may provide no true security at all. This irony is perfectly understandable, however, considering the size and narrow focus of the small fraction of society participating in decisions concerning their design and strategic roles. Military technical decision-making may be the most striking example of a vital area in which the public has been disenfranchised by technical illiteracy, but the future promises more if trends toward that illiteracy are not reversed. A fundamental kind of quality control, that of bringing to bear as much as possible of the wisdom of broad human values and capabilities, will thus be lost.

Quality in science must come, then, from a juxtaposition of favorable institutional arrangements. The scientific community itself must be willing to invest some of its time and talent in self-governance. It must, however, also be willing to accept the integration of other social standards with its own values. It must, in fact, at times call attention to the limits of wisdom, lest an all-too-eager lay society hope for simple or clear answers to questions that have no technical solution. The best of these efforts will be ineffective, however, if the general public is as technologically illiterate as current U.S. survey trends would lead us to expect. For its own interest, as well as that of the public with whom it must communicate, the scientific community must place high priority on efforts to raise the scientific sophistication of the entire society. In the long run, only when the society as a whole can participate in an informed manner can we expect wise decisions concerning the quality and direction of science.

Secrecy and Openness in Science: Ethical Considerations

Sissela Bok

Denunciation of secrecy is ritualistic in modern science. Precisely because the conflicts that secrecy raises for scientists are so strong, their declarations against it are, in part, efforts at conjuring away its power. Unlike other professionals such as lawyers or government officials, modern scientists have never staked out a rationale justifying even limited practices of secrecy. They have held free and open communication to be the most important requirement for their work. Secret practices, although common, have been furtive, and viewed with intense suspicion.

This essay considers the nature of the scientists' ritual repudiation of secrecy, and its function as a deterrent to careful inquiry into the ethical problems that choices between secrecy and openness present. In seeking to disentangle and evaluate the different justifications for certain practices of scientific secrecy, the essay will, I hope, contribute to the study of the effects of those practices on the choices scientists make, and on the quality of scientific research.

Modern scientists have directed their disavowals of secrecy, first of all, against the ancient and powerful doctrine of esotericism in science: the view that only the elect can penetrate to the mysteries of science, and that to open up the scientific process would destroy it, along with the hopes of its practitioners. This view is expressed in the Hippocratic Corpus as follows:

Things however that are holy are revealed only to men who are holy. The profane may not learn them until they have been initiated into the mysteries of science.[1]

In addition to rejecting such an ideal, disclaimers of secrecy in modern science are also meant to assuage the suspicion that scientists might use their knowledge for profiteering and fraud. The public's fear and suspicion of such abuses of secrecy has been especially strong with respect to medicine. In 1769, Samuel Bard advised physicians:

Do not pretend to Secrets, Panaceas, and Nostrums, they are illiberal, dishonest, and inconsistent with your Characters as Gentlemen and Physicians, and with your Duty as Men—for if you are possessed of any valuable remedy, it is undoubtedly your Duty to divulge it, that as many as possible may reap the Benefit of it; and if not (which is generally the Case) you are propagating a Falsehood and imposing upon Mankind.[2]

And the French *Encyclopédie*, published four years earlier, defined a medical secret as a remedy for which the preparation is kept secret in order to raise its potency and price. The combination of rapacity among dissembling physicians and gullibility and hope among patients no matter how near death, the authors concluded, will prolong such age-old practices for as long as the world lasts; it will triumph over the strongest and the clearest evidence; to be surprised thereby, one would have to be *"bien peu philosophe."*[3]

No matter how often denounced, both the pressure for exclusivity and the incentives for profiteering, monopoly, and fraud nevertheless continue in modern science in new guises. In addition, scientists are beset with new pressures for secrecy even as scrutiny of their activities is intensified. The impact of government, business, and military policies of secrecy today reaches into most scientific undertakings, and pits demands for secrecy against adherence to the fundamental norms held to govern science.

Foremost among these norms, according to many observers, is that of open and free sharing of scientific information. It holds that scientists ought not to withhold information but should publish it within a reasonable time, and that their freedom to do so ought not to be limited by any outside institution. Although they may keep confidential those matters pertaining to privacy (of human subjects in experimentation, for example) or ongoing research, they should claim no rights of ownership to what they have discovered or invented, once it is in the public arena. "Secrecy," according to Robert Merton, "is the antithesis of this norm; full and open communication its enactment."[4] Even in the midst of World War II, as scientists around the world were working under conditions of unaccustomed secrecy, the journal *Nature* published a Declaration of Scientific Principles; the seventh of these Principles held that the pursuit of scientific inquiry demands complete intellectual freedom and unrestricted international exchange.[5]

Another norm held forth to scientists is that which Robert Merton has called disinterestedness. It prescribes that scientists, whatever their personal motives, should have the advancement of scientific knowledge as their primary concern.[6] This norm, too, suggests openness, since it is precisely a lack of disinterestedness that leads to secrecy in the practices envisaged by the *Encyclopédie* and others who caution against secret scientific research.

These norms are defended, however, for reasons deeper than the need to combat exclusivism, profiteering scientists, spurious claims to achievements, or arbitrary government demands. The felt need to take a stand against secrecy also springs from concern for what is most central to the scientific enterprise itself: from a recognition of the damage that secrecy can do to thinking, to creativity, and thus to every form of scientific inquiry. Because secrecy limits feedback and restricts the flow of knowledge, it hampers the scientist's capacity to correct estimates according to new information, to see connections, to take unexpected leaps of thought. And secrecy is expensive in that it fosters needless duplication of efforts, postpones the discovery of errors, and leaves the mediocre without criticism and peer review. Secrecy, therefore, can cut into the quality of research and slow scientific momentum.

For these reasons, secrecy affects *all* reasoning and creativity, quite apart from its susceptibility to every form of abuse and pathological excess. But the damage is perhaps especially noticeable in science because of its reliance on reasoning and creativity, and most noticeable of all among those scientists who work in large institutions where secrecy is enforced by surveillance, security clearances, and other methods apt to spread, debilitate participants, and invite abuse.

Scientists are now, as never before, pressed from opposite directions: more openness is required of them, not only by peers but also by the public, yet greater secrecy is pressed upon them at the same time, in addition to their personal incentives for concealment. Edward Shils described this conflict during the McCarthy era:

Science, which for so long had lived apart and to itself, has increasingly in the present century shown how much it can contribute to technology, welfare, and defense ... Under pressure to divulge and not to divulge, science and scientist have fallen into a tormenting cross fire. Offering the quintessence of the most salvationary secrets, scientists have, at the same time and for the same reason, been drawn into the arena of publicity.[7]

Caught in this cross fire, many scientists have fought against secrecy, especially when imposed from the outside. In the 1950s, a number of American scientists stood up in defense of the freedom to do research, and combated the arbitrarily imposed loyalty oaths and security restrictions that affected not only the individuals accused of disloyalty but also, in their view, the conduct of science in its own right. Thus, seventeen American Nobel prize winners, responding in 1959 to a letter from Senator Thomas Henning, agreed (with one exception) that free exchange of information was the life-blood of scientific progress, and that restrictions of this flow were either foolish or destructive.[8] Wartime conditions or extremely sensitive research might require special restrictions, they held, but widespread secrecy was so debilitating that it should be resisted at all costs.

Other scientists were more accommodating. Some made *ad hoc* responses to the demands for secrecy without trying to sort out the ethical issues. Some entered into the spirit of secrecy without reservation, believing that the threats to national security overshadowed concerns for openness in science. Still others took refuge in the claim that research should be approached in an objective, "value-free" manner, and thus succeeded in repressing questions of moral choice in their own lives as scientists as well.

Aside from such issues of national security, the debate over secrecy in science has been much less articulate. Two barriers have impeded it. In the first place, the ritualistic invocation of the norms of openness and disinterestedness in science has served to defuse efforts to analyze the conflict of these norms with secrecy. It is all too easy to slide from the normative to the descriptive in such invocations: to begin to take for granted that what scientists are thought to value must be characteristic of how they act. Such a slide is a form of rationalization that helps keep potential conflicts over secrecy out of view.

A second form of rationalization guards a more limited territory. Many argue that, while difficult issues of secrecy do arise for scientists working in private industry or in the development of technology, this is not the case for scientists in the nonprofit sector, doing research in government agencies or in universities. Only in the commercial arena, they hold, are there serious questions of secrecy; in nonprofit science, openness is still the practice. Although such work may at times require temporary protection of ongoing research, it is meant to be published as soon as possible, and, thus, does not incur the long-term secrecy sometimes imposed on technological work, with all the attendant ethical problems.

The distinction between the nonprofit and the commercial (as between science and technology and between the pure and the applied) is doubtless important; but where ethical problems are concerned, it functions as a smoke screen. Such distinctions obscure the perception of how such problems, including those of secrecy, arise for all scientists.

On the contrary, the lines between ethical issues in commercial and nonprofit science are growing more blurred each day, as the same persons participate in both arenas, and as the results of the most theoretical research find swift, often unexpected application. In the temporary secrecy sought for ongoing research, as well as in longer-term secrecy sought for certain lines of inquiry and for the results of some types of research, scientists are encountering perplexing questions of choice and line-drawing.

Temporary Protection of Ongoing Research

Modern scientists work under conditions of heightened competition, when compared to the situation of their predecessors. On any one frontier of research, many scientists will be at work on very similar projects, often competing for limited resources. Moreover, the level of specialization is such that the researchers cannot easily "shift gears" and turn to a different field. Gaston, in a study of competition in science, compares it to "a race between runners on the same track and over the same distance at the same time."[9] In such a race, only winners are rewarded, and "the researcher who presents what is already known to the community may lose all or most of the recognition that would normally accrue to him."

The desire to be first, to have priority, and to achieve recognition can lead to temporary secrecy about ongoing research. This secrecy is not meant to be permanent; on the contrary, the goal is the most rapid possible publication of results. The only question is that of timing. In order to avoid the risk of being "anticipated" if a competitor borrows and uses ideas or research data and publishes the letter first, investigators may engage in a variety of stratagems. In

The Double Helix, for example, James Watson has described some of the efforts in the race to discover the structure of DNA, to keep competitors (especially Linus Pauling) selectively ignorant of the progress, and told how he and Francis Crick used publication deadlines and waiting periods to protect and reveal when most needed.[10] Horace Judson reports an interview with Max Delbrück that brings out another aspect of the conflict. Watson had written to Delbrück describing the team's new model, adding in a postscript that Delbrück should not mention the letter to Linus Pauling. This caused Delbrück to feel he had a dilemma, which he resolved in favor of openness:

He wanted to tell others in the laboratory about it, he knew Pauling was eager to hear, and anyway he hated secrecy in scientific matters. 'So my [Delbrück is speaking] first reaction was to call up Linus Pauling and say, I have the news, come on over.'[11]

Prizes and personal advancement are important factors in such competitions; for although the knowledge achieved cannot become property in its own right, those who discover it achieve new means *to* property. But they are far from the only factors. One cannot read Watson's book without sensing, also, the excitement of the race in its own right. Others describe the reverse: an oppressive sense of having to compete, of being able only to hint at progress, while worrying about the effect on their research funding if their findings are "borrowed."[12] For many scientists, the fate of an entire research project, involving a number of collaborators, depends on grants for further research—grants which, in turn, depend on a demonstration of originality and importance in research.

The desire for priority undoubtedly fuels innovation, and drives scientists to tolerate frustration, drudgery, and setbacks that might otherwise prove overpowering. But this desire is often concealed since it conflicts with the norms of disinterestedness and of sharing. Robert Merton, in studying this conflict, worked out the following rule: whenever a scientist claims, in an autobiography or a biography, to have little or no interest in priority, there is a reasonably good chance that, not many pages later, the reader will find him deeply embroiled in a battle over priority.[13]

Even when research is going badly, scientists may want secrecy—not so much to achieve priority in the near future as to prevent complete failure in the eyes of outsiders. Thus, one high-energy physicist explains that he is reluctant to make known the fact that an experiment is not going well "because it might get thrown off the accelerator."[14]

Quite apart from the desire for priority, and the fear of being anticipated or of having one's project terminated, scientists may also keep their unfinished work to themselves out of a sense of *reserve*. Indeed, the most meticulous and scrupulous scientists may hesitate the longest before making data public, because they want to refrain from announcement until certain of their results.[15] At issue is the personal claim to control over one's plans and unfinished projects, and over deciding when to reveal them. Unlike the desire for priority, which takes a special and possibly extreme form in science, such reserve has close parallels in all creative work. Many processes of research and creation require a chance to mature, to be tested and varied, without publicity. It is in the nature of such efforts to be tentative at first. Creativity requires freedom to do the unexpected, to risk failure, to pursue what to others might seem farfetched and even pointless.

Whenever exposure of unfinished work is required, the nature of the work itself changes subtly. Notes kept for a possible public are different from purely private notes; just as diaries intended for publication differ from those meant to be kept secret. Outlines, drafts, and diagrams that could be seen by others are not only more elaborate, but are also often efforts at legitimization of the undertaking, and may therefore slip into a certain hypocrisy. It is possible, also, that such documents may take on a special rigidity once they are open to inspection, that other lines of inquiry may seem more sealed off, and the fluidity of the creative process may be lost.

Such developments occur even when no general public is considered, but when work is a team effort. They are even more likely to be strong when the scientific work is funded by government or foundations and, thus, supervised by outsiders from the outset. In each case, the disadvantages of this loss must be weighed against the support received.

Neither the desire for priority nor what I have called "reserve" are ethically problematic in their own right. They rely, in many cases, on legitimate claims to control over plans or property resulting from the investment of resources. To try to wipe out such control would be an intolerable invasion of privacy, and might, in addition, undercut the incentives for creativity and for the significant personal sacrifices scientists often undergo. To the extent that scientists fear theft and anticipation of their ideas, secrecy counteracts those fears and allows leeway for undisturbed research.[16]

When do practices of secrecy in science become problematic? When fraud is uncovered, the scientific community generally deals with it severely. But lesser forms of misrepresentation are harder to pinpoint, and are at times brought about by the conditions of scientific work. The current system of financing research, for example, encourages misrepresentation in the interest of keeping unfinished work from undesired disclosure. Scientists who seek financial support for their work—and that includes nearly all—are now required to describe in some detail what they plan to do if they receive a grant. This process exposes their research in two ways: it opens research plans to potential competitors on the reviewing boards; and it disturbs the privacy of unfinished projects. Yet granting agencies must have some information to go on if they are not to give out money haphazardly or on the basis of favoritism, or only reward past achievement, to the disadvantage of the young or the unknown scientist.

Many scientists acknowledge that, as a result, much vagueness and even falsity beset the requests submitted to funding agencies. Some may describe that portion of their work already finished; others include inaccuracies to throw competitors off the correct track, should they attempt to use the data.

No one should have to resort to deception in order to protect work from plagiarism or premature disclosure. When such motives seem to necessitate fraud, it is often the system itself that must be changed. We might ask, therefore, just how much information funding agencies ought to request from applicants. Without giving up the benefit of peer review, are there alternative ways of making sure that money goes to those who most deserve it? With funding, as with letters of recommendation, many people are now caught up in a vast system of mutual suspicion and misrepresentation that is a disadvantage for all. Asking for too much information, or for the wrong kind of information, merely invites deception as researchers try to protect their work, and this, in turn, facilitates fraud more generally to cut corners and to remain afloat. False and overblown claims create the temptation to live up to these claims at all costs, and encourage imitation and retaliation by those who would otherwise have little incentive to misrepresent their own research or to suspect that of others.

Some have argued that no further limitations are needed: that, so long as fraudulent means are excluded, scientists ought to be free to decide when to keep ongoing research secret, when to publish the results, and how to weigh the advantages of secrecy and openness. Such an argument sounds simple and clear-cut; and it is consonant with the view that scientific work is intellectual property until it enters the public domain. But a rule of this kind is inadequate. Ongoing work is not merely intellectual property when it affects others or requires outside funding. Further limits on degrees of secrecy *and* openness are needed to protect certain kinds of vulnerable ongoing research as well.

The vulnerability of ongoing research to premature publication is especially common in "double-blind" studies with human subjects, in which neither subjects nor investigators know which subjects are receiving the experimental treatment. Such studies require strict concealment. Otherwise it might be impossible to tell whether a subject's responses are attributable to the treatment or to suggestion; investigators might be biased in assigning subjects to groups receiving what they believe to be the most promising treatment; and the results might be interpreted in a biased way.

This vulnerability is methodological. Ideally, the studies should be allowed to run until the hypothesis under investigation is either confirmed or rejected; and they should be conducted with all the necessary concealment. Several circumstances create pressure for openness, however. It may develop in the course of the study that one treatment presents greater risks to participants than another or than a placebo treatment. If so, the safety of the subjects, and their right to know about the risks they run, outweigh the benefits of continued concealment. However, it is hard to know just when in the course of the experiment to conclude that such risks exist. For this reason, the so-called "stopping problem" is the hardest one investigators face. Other pressures may affect the decision about when the experiment has produced valid results of any kind. If investigators lack sufficient training in statistics, they may not understand the length of time and the numbers of subjects required to arrive at conclusive evidence, and therefore rush into print with temporary results. The result of such action can be premature public interest in new drugs or techniques before they have been validated by peers—or a corresponding loss of confidence in existing therapies that are falsely believed to have been proved useless or harmful.

A second aspect of the vulnerability of ongoing research creates conflicts about secrecy for substantive rather than methodological reasons. What is feared, in such cases, is not that experiments will collapse or lose their validity from premature disclosure, but rather that there will be outside criticism of the research. As soon as a study places human subjects or bystanders at risk, it can no longer be claimed to be the investigator's intellectual property, to be revealed or kept secret as he chooses. The history of scientific experimentation is burdened with many examples of studies conducted without the knowledge, much less the consent, of those at risk. In response, guidelines and regulations now seek to make quite clear what information must be provided before a study is undertaken, and to whom; but the difficulties of regulation are great whenever those at risk cannot be identified, or are very numerous. These difficulties are compounded when research is undertaken abroad—as when scientists test contraceptive implants with unknown risks on women in Asia or Latin America, or when a nation tests nuclear devices far from its own shores and its major population centers. The risks may then be very differently understood and assessed on all sides—those who run the greatest risks may be kept most completely in the dark, or have insufficient power to impose any regulation on the investigators.

Other aspects of research, however, may require greater, not lesser secrecy than investigators are prepared to offer. Experiments may be conducted on groups of people, such as alcoholics or the mentally ill, for whom confidentiality is especially important. The information gained in such studies (and identification of individuals as members of such groups) is, once again, hardly to be considered the property of the investigators, to reveal or keep back as they choose. The same is true of all investigations in which questions are asked about matters society regards as private. At present there are many obstacles to the careful protection of confidential, personal information, yet research subjects are rarely informed about them.

For all these reasons, the simple rule that investigators should decide how much secrecy they desire until the time of publication of results is inadequate. Such a rule assumes that the choice does not pose risks—either to participants or to the project—that require others to participate in the decision-making. At times such an assumption may be warranted—for instance in historical or literary research—but in most scientific research, it is not.

Secrecy and the Results of Research

Scientists are also coming under increasing pressure to exercise long-term secrecy over certain research results, and over lines of inquiry capable of achieving certain kinds of results. The pressures come primarily from two directions—business and government. What boundaries might scientists want to draw now with respect to these pressures? Should scientists accommodate to trade or military secrecy less easily than others? Is there something inherent in the role of the scientist which clashes with being used—or using oneself—as a tool for the purposes of commerce or the military?

Trade secrecy envelops not only scientists who work for private business but also many who are based in government and the universities. Companies in many fields eagerly invest large sums in the services of university researchers, asking for guarantees of secrecy in return. Colleagues on the same university faculty may, as a result, find it difficult to share information as openly as before. At scientific meetings, disputes over secrecy are erupting, and in some organizations, actions are being considered which could censure or expel members who use their business obligations as shields to avoid participation in the usual sharing and discussion of new advances. Legal action—such as the suit between the University of California and the pharmaceutical firm Hoffman-La Roche over patent rights to genetic information for the synthesis of interferon—threatens to undercut further the informal exchanges among scientists.

The distinction between pure and applied science, like the related but not identical boundary between nonprofit and commercial scientific work, may be useful for many purposes; but it cannot serve to draw lines of an ethical nature. It is simply not the case that the ethical problems of secrecy somehow come up only in applied or commercial work. Nor can such work be insulated from the other kind. Many scientists engage in both, sometimes in the course of the same day and in the same laboratories. It is absurd to think that questions of how secrecy affects their lives, their treatment of colleagues, their regard for gain, only arise in one of the two contexts. The step from the pure to the applied is often breathtakingly short; considerations of trade secrecy now hover near the most abstract lines of inquiry.

Some ask what is wrong with the role of the scientist as entrepreneur. Why should a biologist, they ask, not use his talents for profit, when economists and artists and inventors have long done so? If secrecy is required to derive the fullest benefits from a scientist's work, why should he hesitate more than others to impose it?

In answering these questions, it is important to stress, first of all, that every entrepreneur, inside as well as outside the academy, in science as well as in economics or architecture, has reason to be cautious about becoming entangled in trade secrecy. True, there are justified forms of such secrecy; but they have vast and debilitating excesses, and many problematic aspects. When trade secrecy leads to personal bondage of either the entrepreneur or employees; when it tempts the entrepreneur to use fraudulent means; when it damages relations with colleagues; when it conceals discoveries or information needed by society (e.g., the discovery of insulin); when it covers up illegal acts or the dangers of a product or process, as in the concealment of harmful side-effects of drugs; and when it is so extensive that it hampers the free communication about large areas of potential development—in such cases, trade secrecy should be at least as great a cause for concern among scientists as among all others.

If we acknowledge that all persons have reason to be cautious, do we also want to go further and say that scientists who work in nonprofit institutions have added reasons to refrain from practices that require them to participate in trade secrecy? Are they different, in this respect, from company-based inventors or freelance writers?

I believe they are. Having been given special privileges and possessing, collectively, unprecedented power, these scientists must assume special constraints. Their work draws upon public funds, and upon the trust and goodwill without which the scientific enterprise would falter. They have an obligation, not only to the public but also to colleagues and students, not to impede the flow of information for private gain. Given this obligation, scientists ought to maintain standards of openness and of disinterestedness that go beyond the caution with respect to trade secrecy that I have recommended for all entrepreneurs.

To argue that scientists have such an added obligation is not, however, to hold that trade secrecy can never be justified in their work. At times, it may be required for purposes of speeding the transfer of technology; at times it may be so innocuous as not to matter. But, in general, special constraints are needed and legitimate in science, over and above the more general caution that trade secrecy should always inspire.

A similar redoubling of caution is required with respect to *military secrecy*. Like trade secrecy, it is justified at times, but carries immense risks of spreading, of creating bondage, of shielding incompetence and corruption, and of delaying advances in knowledge. Like trade secrecy, it removes what is concealed from criticism and feedback, and thus opens the door to abuse; and it is equally capable of inviting imitation and retaliation of every kind. Here, too, scientists have an added responsibility to exercise restraint.

In the last few years, a controversy has arisen in the field of cryptography[17] that shows how difficult it can be to weigh the arguments for and against scientific involvement with military secrecy—and how hard they are to separate from concerns of trade secrecy. Mathematicians and computer scientists have recently made substantive advances in designing virtually unbreakable methods of encoding information. Many of the new encryption schemes would require years, perhaps decades of computer time to decode and these schemes make the codes effectively unbreakable.

As research on such problems began to yield interesting and usable results, the mathematicians came close to the "born secret" region of classified research—in some cases without anticipating that their research would prompt government interference. As they began to publish their results, the National Security Agency (NSA) became increasingly concerned, breaking a 25-year tradition of public silence in order to indicate the need for discussion of what to do about this form of research. The NSA cautioned the mathematicians and, in some instances, made efforts to take over the funding for some of their projects. Researchers rejected these efforts, fearing that a project funded by the National Security Agency could also be censored by it, or classified at its request.

Although the mathematicians have fought these measures in the name of academic freedom, one group of researchers in cryptography did agree to a two-year trial period, beginning in 1980, of voluntary prior restraint. Both researchers and journal editors will submit papers on the relevant topics to the National Security Agency for review. But they have not agreed to suppress papers at the request of the Agency, nor has the Agency explained how it intends to regulate the research—whether through censorship, legal means, or other forms of prohibition.[18]

The concern with such research is understandable. An NSA spokesman argued that secrecy is necessary for research on the making and breaking of codes because, first, other nations could learn that their codes are insecure and could adopt new codes that the NSA cannot break, and second, the research "may enable anyone to make unbreakable codes" and thus interfere with the gathering of intelligence.[19] The United States would also obviously like to be sole beneficiary of such research.

In the case of cryptography research, military secrecy conflicts with academic immunity from outside restrictions and with the norm of open communication among scientists. But another interest is also at odds with the proposed restrictions. In both the private and the public sectors, electronic funds transfer and electronic mail systems make the transmission of personal and financial information highly vulnerable to accidental changes, theft, abuse, and destruction. As a result, banks and other organizations have a strong interest in securely encoding the data transmitted. The market for research on cryptology, therefore, has expanded beyond purely military interests. In both the military and the commercial uses, the purposes are triply concerned with secrecy. The conflicts have to do with the degree of secrecy or openness considered legitimate in the transmission of trade secrets or of military secrets, and with the secrecy needed to hide the possibility of such transmission from outsiders and adversaries.

Academics, faced with this interweaving of separate questions of secrecy, find themselves in a double bind. If, for example, a mathematician is ordinarily in favor of free and open exchanges of information, should he communicate openly research results that could allow more complete secrecy in military and business establishments—including those that are corrupt or violent? He confronts, here, in unpredictable and paradoxical ways, the tensions associated by Joseph Cade with all secrecy in science, arising from the "simultaneous requirement to promote the flow of thought between some sets of minds yet impede it between others."[20]

The difficulty of predicting the results of opting either for secrecy or for openness in this double bind is striking. It is partly a matter of strategy: those who have specific dangers in mind often see good reason not to draw attention to them. But, in addition, the difficulty of prediction stems from our inability to know how things would change if unbreakable codes were available, either to the military alone, or to all who wished to use them. Although scientists do recognize the dangers in pursuit of and publicity about certain lines of research (as with cheap, easy-to-assemble means of mass destruction), the risks from publicity about cryptographic research are less direct, and are at least partially offset by the advantages sought by commercial firms and banks. The greater the power of a collectivity to inflict harm, the more serious the risks of its having means of secrecy that exclude it from accountability. On the other hand, if these means were widespread, they would also provide some protection *against* the most dangerous and invasive groups.

One result of widespread use of unbreakable codes might well be greater vulnerability of the human beings who initiate or receive the messages. As the process of transmission becomes more secure, then more attention will be given to senders and receivers of messages. One would therefore expect an increase in espionage (both industrial and military), bribery, extortion, and other means of probing. To the extent that academics develop such forms of transmission, they could become the targets of and possible participants in these practices.

In considering how mathematicians should respond to pressures to release or keep secret their work on cryptography, however, one must also ask whether preventing the spread of such knowledge is even possible. Can it be kept secure—especially from those most eager to make use of it? Articles have already been published on the subject, alerting readers to the possibility of undertaking further research. Scientists also routinely share such knowledge long before publication by means of pre-prints sent to colleagues around the world. And the voluntary controls being tried will work only so long as *all* researchers agree to participate—an unlikely prospect since the work is now going on in a number of countries and the commercial rewards for the results of such research promise to be high.

This is merely one example of how commercial pressures for secrecy intertwine and sometimes conflict with military pressures. Military secrecy now concerns much more than traditional military matters such as troop movements and defense installations; its focus is primarily technological. As research becomes more and more indispensable to each country's security, conflicts over whether and how to guard it against surveillance and espionage intensify.

The sheer number of scientists touched by this development is a factor in the changed scientific environment. In the industrialized nations, over one-third of the budget for research and development is now allocated to military or space or nuclear studies.[21] Policies of secrecy affect much of this work. If these vast sums, and the scientific talent that they support, could be reallocated even in part to the problems of malnutrition, population, health, environmental deterioration, and energy supplies, the world outlook would change dramatically. If the bulk of research and development support could be redirected to serve human needs, a heavy burden would be removed from the shoulders of scientists.

In considering the quality of scientific research, we must not ignore the effects of current research patterns and the consequent secrecy imposed on scientists. Many have testified to the degrading nature of having to participate in systems of classification, having to swear loyalty oaths, having to fear espionage, or being suspected when traveling or conversing with foreigners.

Scientists cannot abdicate the responsibility for confronting these matters, both jointly as a community of interest and individually from the point of view of how they wish to lead their own lives. How can secrecy be contained or eliminated when unwarranted, and yet strengthened in the narrow regions where it is needed?[22] Many individual scientists and scientific organizations have called for inquiries into the questions of responsibility and choice raised by the growing conflicts over secrecy. Collectively, however, scientists have failed to pursue such inquiries, and to arrive at careful distinctions with respect to what they are asked to do. In the absence of forceful efforts to seek a reasoned response to these conflicts, new practices of secrecy may gain such a strong foothold that they affect the momentum and quality of scientific research in ways difficult to reverse.

Notes

1. Hippocrates, *Law*, translated by W.H.S. Jones, Volume 2 (Cambridge, MA: Harvard University Press, 1923), p. 265.

2. Samuel Bard, *Discourse upon the Duties of a Physician*, excerpted in *Ethics in Medicine*, Stanley Reiser, Arthur Dyck, and William Curran, eds. (Cambridge, MA: The MIT Press, 1977), p. 17.

3. *Encyclopédie ou Dictionnaire Raisonné des Sciences des Arts et Metiers*, Vol. XIV (Neufchatel: Samuel Faulche & Compagnie, 1765), pp. 562–563.

4. See Robert Merton, "The Normative Structure of Science," in *The Sociology of Science* (Chicago, IL: University of Chicago Press, 1973), pp. 267–278. See also Robert Merton, "Priorities in Scientific Discovery: A Chapter in the Sociology of Science," in Bernard Barber and Walter Hirsch, eds., *The Sociology of Science* (New York: The Free Press, 1962), pp. 447–485; John T.

Edsall, *Scientific Freedom and Responsibility* (Washington, DC: American Association for the Advancement of Science, 1975); Ward Pigman and Emmett B. Carmichael, "An Ethical Code for Scientists," 111 *Science* (16 June 1950): 643–647; André Cournand, "The Code of The Scientist and Its Relation to Ethics," 198 *Science* (18 November 1977): 699–705; and "Principles of Scientific Freedom and Responsibility" (draft, 1979), available at the offices of the Committee on Scientific Freedom and Responsibility, American Association for the Advancement of Science, Washington, DC.

5. "The Commonwealth of Science," 148 *Nature* 3753 (4 October 1941): 393.

6. Robert Merton, articles cited in Note 4 above. Mitroff has argued, however, that the norms set forth by Merton have counternorms in scientific work; see Ian I. Mitroff, *The Subjective Side of Science* (New York: Elsevier, 1974). Mulkay sees such norms as part of the ideology through which scientists present themselves to the public; see Michael J. Mulkay, "Norms and Ideology in Science," 15 *Social Science Information* 4/6 (1976): 637–656.

7. Edward Shils, *The Torment of Secrecy* (New York: The Free Press, 1956), pp. 46–47. See also Joseph A. Cade, "Aspects of Secrecy in Science," XXI *Impact of Science on Society* (2 November 1971): 181–190; and Don K. Price, "Security and Publicity Risks," in *Government and Science* (New York: Oxford University Press, 1962), Chapter IV.

8. "Views on Secrecy by American Nobel Prize-Winners," 130 *Science* (10 July 1959): 85–86. See also Shils, *op. cit.*; Francis E. Rourke, *Secrecy and Publicity* (Baltimore: Johns Hopkins University Press, 1961); AAAS Committee Report, "Civil Liberties of Scientists," 110 *Science* (19 August 1949): 177–179; Lloyd V. Berkner, "Secrecy and Scientific Progress," 123 *Science* (May 1956): 783–786; Edward U. Condon, "Science, Secrecy, Security," 200 *Harper's Magazine* (February 1950): 58–63; and I.I. Rabi, *Science: The Center of Culture* (New York: The World Publishing Company, 1970), pp. 101–110.

9. Jerry Gaston, *Originality and Competition in Science* (Chicago, IL: The University of Chicago Press, 1973), p. 74. See also John Ziman, *Public Knowledge* (Cambridge, MA: Cambridge University Press, 1968), pp. 96–98, and Chapter 6; and Warren O. Hagstrom, *The Scientific Community* (New York: Basic Books, Inc., 1965).

10. James D. Watson, *The Double Helix* (New York: Atheneum, 1968).

11. Horace F. Judson, *The Eighth Day of Creation* (New York: Simon & Schuster, 1979), p. 177.

12. See, for instance, June Goodfield, *An Imagined World: A Story of Scientific Discovery* (New York: Harper and Row, 1981); and J.D. Bernal, *The Social Function of Science* (New York: Macmillan, 1939), pp. 107–108. For a discussion of who controls scientific information, see Dorothy Nelkin, "Intellectual Property: The Control of Scientific Information," paper prepared

for AAAS Committee on Scientific Freedom and Responsibility for its annual meeting, January 1982.

13. Robert Merton, "The Ambivalence of Scientists," in Norman Kaplan, ed., *Science and Society* (Chicago, IL: Rand McNally, 1965), p. 116.

14. Gaston, *op. cit.*, p. 124.

15. *Ibid.*, p. 123. See also Spencer R. Weart, "Scientists with a Secret," *Physics Today* (29 February 1976): 23–30; and Ziman, *op. cit.*, p. 97.

16. Ian I. Mitroff argues that secrecy can thereby further the progress of science, in *The Subjective Side of Science, op. cit.*, p. 76. For a study of motivations and pressures among scientists, see Bruno Latour and Steve Woolgar, *Laboratory Life* (Beverly Hills, CA: Sage Publications, 1979).

17. Deborah Shapley, "Intelligence Agency Chief Seeks 'Dialogue' With Academics," 202 *Science* (27 October 1978): 407–410; Gina Bari Kolata, "Cryptography: A New Clash Between Academic Freedom and National Security," 209 *Science* (29 August 1980): 995–996. For a study of the history and forms of cryptology, see David Kahn, *The Code-Breakers: The Story of Secret Writing* (New York: Macmillan Publishing Company, Inc., 1967).

18. Gina Bari Kolata, "Prior Restraints Recommended," 211 *Science* (20 February 1981): 797. See also Gina Bari Kolata, "MIT Committee Seeks Cryptography Policy," 211 *Science* (13 March 1981): 1139–1140.

19. Gina Bari Kolata, "Prior Restraints on Cryptography Considered," 208 *Science* (27 June 1980): 1442–1443.

20. Joseph Cade, *op. cit.*, p. 182.

21. See, for example, *North-South: A Program for Survival* (Cambridge, MA: The MIT Press, 1980), authored under the chairmanship of Willy Brandt, and Colin Norman, *The God That Limps: Science and Technology in the Eighties* (New York: W.W. Norton and Company, 1981), pp. 71–78.

22. I have discussed the question of whether some research should be not only kept secret but also actually declared out of bounds or prohibited, in "Freedom and Risk," pp. 115–128 in Gerald Holton and Robert S. Morison, eds., *Limits of Scientific Inquiry* (New York: W.W. Norton and Company, 1979).

Ethical Issues in the Assessment of Science

Excerpts from a Seminar Series

Editor's Introduction—*From December 1979 through December 1980, a series of faculty seminars were held in Cambridge, Massachusetts, under the auspices of a project on "Ethical Issues in the Assessment of Science: Development and Testing of Indicators of Quality in Science and Technology," funded by the National Science Foundation and the National Endowment for the Humanities. These seminars, led by Gerald Holton, explored a number of different perspectives on science indicators, engaging the interest of historians, philosophers, sociologists, and political scientists, as well as research scientists and science policy experts.*

In many of the seminars, participants expressed concerns about an interesting side issue—the effect on the quality of science, and on the public's assessment of science, of the overt linking of science and politics. The discussions occasionally touched on what could be best described as "separation of powers" issues—where the powers engaged are those of science and the political establishment. How is emergence of this ever-strengthening link affecting the allocation of resources, the choice of research topics, the regulation of scientific inquiry, and public attitudes toward control of science? A sampler of dialogue from the seminars is offered here as a spark for further discussion.

On 1 March 1980, Frank Press (who was at that time Science Advisor to President Jimmy Carter) talked to the seminar group about the types of indicators his office used in recommending policy directions to the President. As the seminar drew to a close, one of the participants, Leo Marx (Professor of American Cultural History, MIT), commented on discussions that had taken place earlier:

Leo Marx At the first meeting of this group, we heard quite a lot about the changing public and Congressional perceptions of science and technology, and the increased skepticism about science's value to the country as a whole.

... As I listen to the discussion this afternoon, I sense a certain paradox here, insofar as the office of Presidential Science Advisor becomes more prominent, powerful, and efficacious, it becomes identified with the apex of political power in this country—the White House—and that might not necessarily be such a great thing for science. If we want the American people to have confidence in the independent contribution of science and technology to the national welfare, and if we are proposing to do long-range planning for science and technology, then it might be better if the plans did not come out of the White House, which, after all, has a very special significance. Perhaps such planning ought to be located where there is more sense of the involvement of areas of society other than the political establishment. ...

... [T]he high public prestige of science and technology, coming out of the Enlightenment, was a result of the belief that it was an independent force for the betterment of human life. To the extent that Big Science is identified with specific goals of this or any other group, class, or nation, it will be hurt by that identification. Around MIT, and elsewhere, many scientists and engineers view with incredulity the reports of increasing public disaffection with science. ... I wonder whether it might be harmful (for the public perception of science) if science were to be identified too closely with specific political programs and policies. That, as against the greater value of the Science Advisor seeing the President every day, there is a sense in which science ought to be separated from the apex of political power. From my point of view, it is scary to see the science establishment so closely identified with political power ... I am not making a case for having the Science Advisor removed from the White House. Rather, I am suggesting that specific constraints or limits might be placed on the Science Advisor and the Office of Science and Technology Policy ...

[As long as that Office is part of the political process] then scientists and engineers should not be surprised if public attitudes toward science and engineering change in direct proportion to change in public attitudes toward government. If the public loses confidence in government, science is going to suffer along with it, because we are equating the two, we are putting the two close together.

These same concerns surfaced again, phrased in slightly different language, in the seminar given on 28 April 1980 by Kenneth Prewitt, President of the Social Science Research Council. During the formal part of his presentation, Dr. Prewitt talked about some of the effects—especially on the conduct of science—of the public perception of the "proximity of science to democratic politics."

Kenneth Prewitt Over the past several decades, there has been gradual and, in some instances, problematic expansion of two things. One is the attentive public for science—that is, there is now a broadened constituency of non-scientists who attend to and often influence science and technological matters. Looked at from the point of view of the mass public, this attentive public is a tiny and highly selective constituency; but looked at from the point of view of practicing scientists, it appears as a large and unpredictable actor . . .

. . . I certainly don't need to repeat to *this* group that matters scientific are now on the national and local public agenda. . . . Science also allows us to do many things that we did not do before, such as modify weather, control population, set nutrition policies, etc. You cannot get as close to the centers of power as science has gotten and not expect issues of science to get on the agenda. In 1975, more than half of the legislation introduced in Congress concerned a scientific or technological topic. Indeed, as you know, some people extend this point to say that the very presence of science in the society is rearranging the political structure: for example, the Lasswellian concept of the "garrison state," or Barry Commoner's idea that allowing development of nuclear power will destroy civilian control of the military, separation of powers, federalism, etc. Substantial political charges are being put at the door of science and technology.

The very proximity of science to democratic politics has two large-scale policy implications. First, matters which are only—or perhaps not at all—issues of science will nevertheless have consequences for attitudes toward science. If I wanted to examine the relationship of public attitudes toward funding for science, . . . I would look at public attitudes toward defense budgets, toward "balance of the budget" issues, or toward women in science. There is a whole series of things about which the public has expressed feelings and concerns that have greater consequence for the health of science than the public's attitudes toward science itself. If we are really interested in the relationship of the mass public to science, then we must look at the way science is being "bounced around" by other things . . .

In some respects, such things [as national budget deliberations or the passage of special legislation] have more important consequences for science than attitudes about whether the public has confidence in the leaders of the scientific community. Once we get that close to the political process, the priorities in science and the levels of funding for science will be influenced by other priorities emerging from the political process.

This does not mean, however, that there are no policies governing science. Guidelines for the protection of human subjects, cost accounting standards, research priorities set by funding levels, regulation of laboratory procedures, and so forth are not trivial in their effect on science.

Policies move in the society from a democratic political culture to a scientific culture. Kenneth Boulding, in his address to the 1980 Annual Meeting of the American Association for the Advancement of Science, stated this very nicely:

Science occupies a social-ecological niche, the boundaries of which are very largely determined by the image of science in the minds of nonscientists, especially those who make decisions about budgets, whether in government, universities or industry.

The science culture is largely self-justifying, believes in its intrinsic worth, believes in human curiosity, believes in the quest for truth and so forth. The democratic political culture is really built upon the concept of promise: "we will deliver if you give us your vote, your allegiance, your taxes—and we will produce." The entire logic of democracy is that a population's support is earned by responsibilities accepted by the regime or the political leadership. The purpose, then, of the democratic political culture is based upon the concept of deliverables. The purpose of the scientific culture is obviously based upon the concepts of intrinsic worth, the quest for truth, and so forth. When policies begin to move from the democratic political culture to the scientific culture, they will, I think, have a very strong theme of utilitarianism.

Also, the scientific culture is based upon concepts of autonomy, peer control, the internalized ethic, the self-directed quest for truth. The political process in the democratic political culture is based on such concepts as accountability or probationary states (you are only there as long as you deliver) and, indeed, support may be withdrawn if a government does not deliver on its promises. If you move from a culture preoccupied with accountability to a culture determined

Ethical Issues

by or structured around the values of autonomy, then you are bound to introduce regulation.

... These two themes—the application of utilitarian criteria and increased regulatory attention to science—constitute the substantive meaning of the frequently repeated phrase "we're renegotiating the contract between science and society." ... the content of that phrase is a function of the juxtaposition of the science culture with the democratic political culture ...

There are a host of reasons why utilitarian criteria are emerging out of the democratic political culture, resting largely upon the concept that you must promise in order to elicit trust, confidence, taxes, and so forth. Allegiance from the public is, from my perspective, highly risky for the scientific culture.

Harvey Brooks (Harvard University) Then what you are saying is that the democratic culture forces scientists to make promises in order to get support and then they get trapped.

Kenneth Prewitt We start with the premise that science is getting closer to power. As it gets closer to power—and since democratic political power is preoccupied with the notion that you must be accountable for what you do—then the translation of such accountability for science will be in the notion of regulation. As I said, that aspect can, of course, be negotiated on terms that will not be as damaging as feared.

But the democratic political culture is not only preoccupied with accountability, it also has an inherent system of promises and benefits and exchange, and that's where I see the risk for science. ...

In the responses to Dr. Prewitt's talk, Dorothy Nelkin (Science, Technology and Society Program, Cornell University) and Alexander Morin (then Director of the Office of Science and Society, National Science Foundation) commented further on how public policy issues influence current public attitudes to science, and Leo Marx remarked upon Prewitt's model for the interplay between science and the public within a democratic system. Each of these discussants appeared to see significant—but still largely undefined— changes taking place in the politicization of science.

Dorothy Nelkin [responding to Kenneth Prewitt] One of the most interesting pieces of survey data I've seen on this topic relates to an issue that you haven't raised. The National Commission for the Protection of Human Subjects of Biomedical and Behavioral Research found that 59% of the general public felt that science as well as technology must be subject to greater public control. This seems to contradict some of your data, but it does coincide with the increased discussion of the accountability of research and also seems to follow from the demands of protest groups concerned about science.

The reasons for such emphasis on accountability include the high cost of research and the fact that more and more technologies appear to be intrusive. In addition, an increasing number of other public issues are being redefined in terms of science and technology. In other words, problems of risk could be defined as problems of corporate power or responsibility, but instead they often are defined as problems inherent to technology itself . . .

An important question, it seems to me, relates to the actual *impact* of public pressure on the research process. I would argue that, contrary to the fears of many scientists, the impact has been minimal. If you look at the Institutional Review Boards, the Ethics Advisory Board, the new Presidential Commission, a systematic pattern seems to prevail. Despite considerable lip service to public participation, these boards remain dominated by scientists. The terms of the debate are structured by the scientific community. The formulation of problems in each area tends to emphasize questions amenable to technical judgment. And the guidelines that have emerged, for example, from the recent *in vitro* fertilization dispute, from the earlier debate on fetal research, and from the Recombinant DNA controversy, have not really imposed very serious limits on scientific inquiry. In a climate where trust in authority is rather low and regulations often substitute for individual responsibility, the scientific community has fared rather well. Indeed, one might argue that the growing concern about science among the attentive public has provided some remarkable opportunities for scientists themselves to shape public opinion, not only about science but also about many other policy areas. Controversial areas of science inevitably engage scientists in disputes and so are the sources of jobs in consulting firms, and positions on commissions, committees, boards, and panels. Scientists have increasingly entered political life as publicists, interveners, consultants, adversaries, and often as *quasi*-representatives of competing interest groups.

The implications of these developments are worth considering. What does this increased involvement of scientists imply for the process of democratic decision-making in key policy areas? Does mass public faith in science justify apolitical decision-making authority in public policy? Do we really want to shift from a political to a more scientific culture? And, ironically, there are implications for science itself—because the more the scientific culture becomes a reference for political choice, the more it's likely to become the focus of political demands and political disputes. . . .

Alexander Morin [responding to Kenneth Prewitt] . . . Implicit in the model [being discussed here] is a limitation of science and technology policy to the very narrow question of how much money is allocated by the Federal Government, or by the economy, to scientific and technical research (in particular, research carried out in the academic environment) and how funds are allocated within that total. If science and technology policy is defined in these narrow terms, then we can almost inevitably conclude that most people are not interested in that issue. It is a trivial proportion of the total Federal budget, if you exclude all of the developmental and applied technology work. I believe that most people will not have an opinion about basic research; it will not loom large as a significant public policy issue in any sense.

What do loom large are the issue-specific decisions about energy, transportation, housing, health, for example. All of these are highly political issues that are being addressed by people who want to participate because of both their special interests and a general interest in public policy. All of these issues have profound effects on the conduct of substantial pieces of the total scientific and technical enterprise; they are reflected in the R&D budgets and in the types and quality of and limitations on research. . . . There is enormous feedback from defense research or from health research into what we would regard as the narrower question of science policy. . . .

My intuitive conclusion is that there is very little direct spill-over from any of these issues—including the populism built into much of the current legislation at Federal, state, and local levels, as well as feelings about participation in decision-making—to questions about the conduct of the scientific and technical enterprise. But there is *enormous* spill-over in the public's perception of how decisions directly affecting its interests ought to be made.

The amount of allocation of the Federal budget that goes to basic research is not one of those issues. What *is* one of those issues is how to vote on a bond issue for either a high-technology or a low-technology sanitation plant located in the district. What *is* involved is a state referendum on weather modification legislation, such as has taken place in Oklahoma. Such issues are not interpreted by those publics as scientific and technical issues. They are interpreted entirely as political issues, as issues related to who has power and how it is organized and how it is exercised and for whose benefit. If that political process is to continue, certain levels of scientific and technical information may be required and the people who control that information may or may not be trusted, depending on who they represent, but it is in a completely political context. Those are not science and technology issues; they are political issues. . . .

Leo Marx [commenting during the discussion of Kenneth Prewitt's talk] I cannot help feeling that part of the reason your survey data seem so reassuring has to do with how the model of science is initially defined, in particular, because of the picture of science used—that is, an opposition between the public or a political system, pressing a utilitarian view, and a community whose interests are in autonomy, truth, disinterestedness. The image suggests a world out there pressing in and threatening to corrupt what appears to be a pristine community. You have said that you thought the protest would, in fact, continue, but that it would not be against science and technology *per se*, rather it would be against the attributes that surround it. Now, that strikes me as a rather crucial point. What are those attributes? I suspect that the attributes reveal connections between science and technology which are very important in the public's perception. If you define science and technology *independent* of those "attributes," you can get a rather misleading picture, because it is precisely those attributes that are part of the real situation. Aren't they, in fact, the connection between science and technology and the system of power? And, if they are, then isn't the public *right* in identifying science with a certain place in the power structure and not with others? If you begin with an initial definition of science which separates it from its historical and social context and say, "Yes, there will be a protest, but it isn't really against science, it's against these extraneous matters," then we must ask if they really are that extraneous.

Afterword

Marcel Chotkowski La Follette

The essays in this book were commissioned with an aim of advancing discussion beyond foot-stamping criticism of prominent indicators of quality in science. The essayists were challenged to scrutinize what Harvey Brooks calls the "inherent tension between simplicity and reality," to propose new approaches to the problem, and to introduce into policy discussions the ethical and value issues that affect the invention, selection, and use of particular indicators of quality in science.

The essayists and seminar commentators approach these topics from several different perspectives. Some, like Lewis Branscomb and Harvey Brooks, report on what actually takes place when indicators have been or are being used for decision-making. Branscomb's description of the decision-points in IBM Corporation's evaluation of researchers, research, and product development affords a window onto a process that could have applications in other settings. Brooks' retrospective assessment of decision-making in four research areas has similar usefulness, for it gives exposure to the processes in university-based work, where the values of the individual investigator or field can influence powerfully the choice of topics, methods, or hypotheses.

Restraints on research also may affect the validity of certain indicators. In her essay on secrecy in science, Sissela Bok provokes many complex questions about the applicability of indicators to newly-classified fields such as cryptography, where the results are important for both military and civilian users of computers and high-speed data processors, but the control of publication (and, hence, of widespread recognition) rests with only one group. Brooks' description of how scientists may regard certain topics as "low-grade," "routine," and "scientifically not very challenging," and therefore "not likely to attract the interest or attention of independent investigators," provides examples of internal restraints that affect the validity of indicators.

Many of the contributors emphasize the importance of the dichotomy between internal and external indicators—for example, "intrinsic and extrinsic value" of research (Branscomb), "inside and outside measures of quality" (Mazlish), or "intrinsic vs. extrinsic indicators" (Brooks). As Mazlish points out, quantitative indicators tend to be concerned with the "inside" image of science, but their predominance should not obscure the importance or significance of "alternative modes of understanding the quality of science, especially its 'outside.' " The three essays by Yankelovich, Prewitt, and Weingart provide the backdrop to these discussions through their often sharply disparate views of (1) how science should "seek thoughtful public judgments about priorities" and (2) how it reacts to social challenges to the legitimacy and direction of research. The writers paint pictures of "science increasingly politicized" (Weingart), of science no longer allowed to "stand aloof" from changing political and social values (Yankelovich), and of the increasing impingement on science policy of "deeply held political values of democratic accountability and public scrutiny" (Prewitt).

Examples of such values underpin the statements by Senator Hatch and Representatives Brown, Fuqua, and Walgren. Because they serve on committees that oversee government-funded or government-regulated scientific and technical research, these members of Congress are representative not only of their voting constituencies but also of the "consumers" of indicators of quality, that is, the decision-makers who look for quantitative and qualitative measures for help in allocating resources or directing research policy at the national level. Brooks' question "Will the science we are generating today be the science wanted and needed to attack the salient problems two decades hence?" puts into words the uncertainty feeding the Congressmen's underlying tone of impatience with scientists. For, despite the fact that policy for science and technology is, by and large, directed by *policy-makers* (who, even if they are working scientists, perceive their role as a double one), the non-scientist's perception is that *scientists* maintain control and, as Hatch forcefully states, must therefore acknowledge a "special responsibility" for directing science.

One thing that these essays—and the seminar discussions preceding them—clearly indicate is that this is a difficult problem. Why? Three reasons spring immediately to mind—although they are by no means the only ones.

First, as a practical matter, measures of "quality" are, by their very nature, difficult to devise. The insufficiency of quantitative measures pushes the development effort; but, when discussion gets to the point of concrete suggestions for a real indicator for real-time use, inevitably we ask what such an indicator would look like. How would it be used? Are there data to feed into it? As the Comptroller General's critique of the *Science Indicators* series states, at the practical level "it is how well the *data* fit and measure a relevant concept that determines the indicator's validity" (emphasis supplied).* Consideration of indicators of quality must be a process in which proposals for new approaches or new measures proceed hand in hand with comprehensive information about existing data sets and/or sources and with proposals for the generation of new data (such as the public attitudes research described in the Yankelovich seminar).

Second, there are a host of "political" reasons, some having to do with real or perceived attitudes within the community of social scientists concerned with science indicators and some having to do with the sensitivity of both the choice of measures and the real or potential conclusions based upon them. Each of the writers refers to this situation, some enigmatically but all cognizant of the significance of the influence.

A third, mitigating factor is the effect of the tugs and pulls of the pertinent constituencies' differing value systems. Conflicting social, economic, or political pressures can make the requirements for a usable indicator at least appear different at the different decision-points. Let me name just a few such constituencies. (a) *Scientists* resist the "unscientific" self-scrutiny implied by formal indicators of quality; they are much more comfortable with informal, intuitive, "instrinsic" indicators guided by the norms of science and closely fitted to the history, demands, sociology, current "success," and morale of each field or research area. If you ask scientists about their research agendas or why they do what they do, one colleague observed, they get defensive because they feel obliged to explain; but if you ask them

---

* Report by the Comptroller General of the United States, *Science Indicators: Improvements Needed in Design, Construction, and Interpretation*, PAD-79-35 (U.S. General Accounting Office, Washington, DC, 25 September 1979), p. 9.

to describe how they made a certain decision, they may talk for hours. (b) *The public* may have its own special set of indicators of quality. Certainly, one criterion is the potential economic and social "value" (positive and negative) of the research product. Another criterion may be the timing of the proposed investment—the public may not always evaluate the same research in the same way. The size of the investment required may also be important—although it is not simply that big is always better or that small is always considered more economical. A small research project that meets few other criteria could be evaluated as a waste of time and resources. For positive evaluation by the public, scientific research also must not conflict with contemporary cultural ideals, myths, or values, and must match the public's expectations of how and when those ideals may be achieved. Positive public assessment may also be significantly affected by the political or economic strength of the groups benefited or harmed by the research. (c) The value systems of *the users* of indicators—in government, in industry management, in Congressional committees—play important (and occasionally conflicting) roles in determining the parameters for new indicators and in directing their use. The user's choices can be governed, among other factors, by political considerations, by availability of resources, and by the "results" the user desires. One of the questions for consideration in the future might be the differing requirements of the various users.

Given that this is a difficult problem (and I have touched on only a few of the difficulties), in what direction do these essays point? Each reader will have her or his own agenda arising from questions raised here, but let me make two observations about some threads apparent in the essays, just to get the discussion started.

(1) The ethical aspects of indicators
The essays clearly show that considerations of the ethical and value aspects of science are germane to the measurement of its quality. Here are just a few questions raised by the essays and seminars: Is one indicator of quality of science the openness of the enterprise to all promising science students, male and female, black and white, Native American and naturalized refugee? Or the availability of research funds to young scientists just starting their careers? How can these be measured? What about measures of scientific freedom—at the level both of the individual researcher and of the field? What

Afterword

about the economic status or job security of young scientists? or the effect of rapid "industrialization" in fields such as genetic engineering? Much of the science indicators discussion has focused on measures for choosing the "right" proposals, the upper 10%, the research with the "greatest potential." The other side of that coin, however, may be an approach that asks how we can assure that proposals (or people) are not rejected for the wrong reasons.

This type of search for underlying ethical assumptions can help to assure the fit of science indicators to both scientific and social reality. As Fuqua and Walgren summarize, "only when the society as a whole can participate [in these choices] in an informed manner, can we expect wise decisions concerning the quality and direction of science."

(2) Indicators and the public

Public opinions about science are usually responded to only when they are hostile, only when there is a crisis or controversy. This is unfortunate. The Public Agenda Foundation research described by Yankelovich*—which, as of this writing, is in the data collection stage—represents one of many possible experimental approaches to measuring the non-scientific public's assessment of scientific quality. Upon observing some of the survey research groups (composed of ordinary citizens discussing scientific research proposals), I was surprised at the participants' apparent unquestioning assumption that they *can* talk about research priorities in a positive way, and that they have a right to discuss such things. The scientific community—and the social scientists and policy-makers concerned with science—have, I believe, long felt that public attitudes represent primarily a negative pressure on science, especially when scientists argue over the public's "right" to limit science or to have certain technical information. The Public Agenda Foundation survey does not introduce false or "let's pretend" controversy into the presentations to the participants; rather they are presented with examples of real dilemmas in resource allocation and research priorities as positive, serious questions. These "average" Americans do not *appear* (and I use that word deliberately

---

* The research is being undertaken by the Public Agenda Foundation, under subcontract to Harvard University. Persons involved with the survey include Daniel Yankelovich, John Doble of the Public Agenda Foundation, Gerald Holton of Harvard University, and the author.

because of the early stage of the survey research) overtly hostile to science *per se,* although they seem to recognize intuitively the politicized nature of science, even of basic science.

The results from the experimental survey—as from the Miller, Prewitt, and Pearson survey for the National Science Foundation*—will undoubtedly raise a host of new research questions. But one observation that I might make now from our preliminary data and from the essays in this book is that serious consideration can and must be given to how to give "weight to the social functions of the scientific enterprise" (Brooks), and how to construct a "more effective decision-making partnership between the public and the scientific community" (Fuqua and Walgren)—and that these are *not* extraneous matters.

*Jon D. Miller, Kenneth Prewitt, and Robert Pearson, *The Attitudes of the U.S. Public Toward Science and Technology* (Chicago, IL: National Opinion Research Center, University of Chicago, July 1980).

Contributors

Sissela Bok is the author of *Lying: Moral Choice in Public and Private Life* (1978) and *Secrets: On the Ethics of Concealment and Revelation* (1982), and has lectured in the Harvard Medical School and the Kennedy School of Government on the subject of ethics.

Lewis Branscomb is Vice President and Chief Scientist, IBM Corporation, Armonk, NY, and the Chairman of the U.S. National Science Board.

Harvey Brooks is Benjamin Peirce Professor of Technology and Public Policy and Professor of Applied Physics, Harvard University, Cambridge, MA.

George E. Brown, Jr., is the U.S. Representative from the 36th Congressional District in the State of California and senior member of the House Committee on Science and Technology.

Don Fuqua is the U.S. Representative from the 2nd Congressional District in the State of Florida and Chairman of the House Committee on Science and Technology.

Stephen Graubard is Editor of *Daedalus* and Professor of History, Brown University, Providence, RI.

Orrin G. Hatch is the U.S. Senator from the State of Utah and Chairman of the Senate Committee on Labor and Human Resources.

Gerald Holton is Mallinckrodt Professor of Physics and Professor of the History of Science, Harvard University, and Visiting Professor in the Program in Science, Technology, and Society, Massachusetts Institute of Technology, Cambridge, MA.

Marcel Chotkowski La Follette is Editor of *Science, Technology, & Human Values* and Research Associate in the Department of Physics, Harvard University, and the Program in Science, Technology, and Society, Massachusetts Institute of Technology, Cambridge, MA.

Contributors

Leo Marx is William R. Kenan Professor of American Cultural History, Program in Science, Technology, and Society, Massachusetts Institute of Technology, Cambridge, MA.

Bruce Mazlish is Professor of History, Department of Humanities, Massachusetts Institute of Technology, Cambridge, MA.

Alexander Morin is Director, Division of Intergovernmental and Public Service Programs, National Science Foundation, Washington, DC.

Robert S. Morison is Visiting Professor of Science and Society, Program in Science, Technology, and Society, Massachusetts Institute of Technology, Cambridge, MA.

Dorothy Nelkin is Associate Professor, Program on Science, Technology, and Society, and Department of City and Regional Planning, Cornell University, Ithaca, NY.

Kenneth Prewitt is President of the Social Science Research Council, New York, NY, and Adjunct Professor of Political Science, Columbia University.

Willis H. Shapley is Public Policy Consultant to the American Association for the Advancement of Science, Washington, DC.

Doug Walgren is the U.S. Representative from the 18th Congressional District in the Commonwealth of Pennsylvania and Chairman of the House Subcommittee on Science, Research, and Technology.

Peter Weingart is Professor of Sociology, University of Bielefeld, Bielefeld, Federal Republic of Germany.

Christopher Wright is the Staff Member for Science Policy and International Development, Carnegie Institution, Washington, DC.

Daniel Yankelovich is Chairman of Yankelovich, Skelly, and White, Inc., and President of The Public Agenda, New York, NY.

Index

Accountability, of government, 152–154
Almond, Gabriel, 87–88
American Academy of Arts and Sciences, 57
American People and Foreign Policy, The, 87
"Anti-science" attitudes, 97. *See also* Public attitudes to science
Aristotle, vii
Autonomy, of professions, 116–117
Auto safety, research on, 11–12

Bard, Samuel, 132
Bauer, Raymond, 50, 55, 57–58, 60–61
Bell, Daniel, 55
Bell Laboratories, x
Benison, Saul, 47
Bentham, Jeremy, 48, 53
Bodian, David, 43
Bok, Sissela, 157
Boulding, Kenneth, 82–84, 86, 152
Branscomb, Lewis, 157–158
Brigham Young University, 119
Brooks, Harvey, 51, 55, 153, 157–158, 162
Brown, George E., Jr., 158
Bush, Vannevar, 3, 38, 123

Cade, Joseph, 144
California, University of, 141
Cancer and the environment, research on, 12–15, 44–45

Carter, Jimmy, 36
Carter Administration, 109, 124
Chemical Abstracts Service, ix
Citation analysis, x, 1, 33–34, 36, 46
Commoner, Barry, 151
Competition in science, 135–136
Comprehensive Environmental Response Compensation and Liability Act of 1980, 30
Comroe, Julius H., Jr., ix, xi
Confidence in technology, 103
Consumer Price Index (CPI), 54
1 Corinthians, 119
Creationism, ix
Crick, Francis, 136
Crime series data, 56
Cryptography, research on, 143–146
Culliton, Barbara, 116

David, Edward, 37
Declaration of Scientific Principles, 132–133
Delbrück, Max, 136
Doble, John, 161
"Double-blind" studies, 139
Double Helix, The, 136
Dripps, Robert D., ix, xi

East-West Center, 21
Economic Indicators, 54
Economic indicators, 1, 2, 33, 54
Economic Report, 54–55

Index

Einstein, Albert, 126–127
Encyclopédie (1765), 132–133
Enders, John, 43
Engineering education, 79–80
"Extra-scientific" social standards, viii
Extrinsic/intrinsic measures, x, 2–4, 69, 74, 81, 158–159
Extrinsic measures, development of, 5

Federal Republic of Germany
　economic growth of, 22
　freedom of research in, 114
Federation of American Scientists, 63
Fermi awards, 9
Finlay, Carlos, 39
Flexner, Simon, 40
Ford, Gerald, 84
Foreign Affairs, 87
Foreign Policy, 87
Frankenstein, 119
Fried, Charles, 3
Friedman, Milton, 123
Fuqua, Don, 158, 161–162

"Garrison state," 151
Gaston, Jerry, 135
General Electric Company, 75
Genetic engineering. *See* Recombinant DNA
Gilder, George, 123
Global 2000 report, 19–20, 120
GNP. *See* Gross National Product
Gorgas, William C., 40
Graubard, Stephen, 110
Greenspan, Alan, 54
Gregory, A. S., 84
Gross National Product (GNP), 1, 33, 54

Hackerman, Norman, 84
Hatch, Orrin G., 158
Health of science, viii, 1, 20, 45, 48, 49–50, 65
Health of the Scientific and Technical Enterprise, The, 49

Henning, Thomas, 134
Hepatitis, research on, 43–44
Hippocratic Corpus, 131
Hoffman-La Roche, 141
Holton, Gerald, 77, 149, 161
Hornig, Donald F., 49

IBM Corporation, 70–78, 157
Innovation, x, 4, 124
Input/output measures, xi, 1, 33, 36
Inside/outside measures, 50–52, 63–65, 158. *See also* Extrinsic/intrinsic measures
Institutionalization of science, 113–118
Institutional review boards, 154
International Energy Agency, 17
International Federation of Institutes for Advanced Study (IFIAS), 21
International Institute for Applied Systems Analysis (IIASA), 20–21
Intrinsic/extrinsic measures. *See* Extrinsic/intrinsic measures
Italy, economic growth of, 22

Japan
　economic growth of, 22–23
　industrial innovation in, 23
Jefferson Lectures, 83
Jencks, Christopher, 46
Josephson superconducting technology, 75
Judson, Horace, 136

Kelvin, Lord (William Thomson), 48, 65
Kennedy, Donald, viii
Kepler, Johannes, 48
Keyworth, George A., viii
Kistiakowsky, George, 63

Lasker, Mary, 35
Latour, Bruno, 148
Lawrence awards, 9
Lawwill, Stanley, Jr., 84
Loweth, Hugh, 36–37

Index 167

McCarthy, Joseph, 133
Marx, Leo, 62, 149, 153, 156
Massachusetts Institute of Technology, Program in Science, Technology, and Society, vii
Mazlish, Bruce, 158
Mazur, Allan, 85, 99
Merton, Robert K., 52–53, 132–133, 136
Michelson-Morley experiment, 69
Michigan, University of, 104
Miller, Jon D., 56, 59, 98, 162
Mitroff, Ian I., 147–148
Moral Majority, 87, 104–105
Morin, Alexander J., 109, 153, 155
Morison, Robert S., 11–12, 63
Morrison, Philip, 63
Mulkay, Michael J., 147

Nader, Ralph, 11, 107
Nation, The, 87
National Academy of Sciences, 63, 124
National defense policy, 129
National Opinion Research Center (NORC), 101
National Research Council, ix
National security, 3, 134
Nature, 132
Nazi government and research, 114
Nelkin, Dorothy, 147, 153
Nobel prizes, as indicator of quality, x, 1, 4, 33, 36, 46
Noguchi, Hideyo, 40
Norms of science, 132–133, 136, 142, 156, 159

O'Connor, Basil, 42
Oresme, Nicole, vii
Organisation for Economic Cooperation and Development, x, 4, 17

Panama Canal Treaty, public attitudes to, 109
Passmore, John, x
Pauling, Linus, 136
Pearson, Robert, 162

Peer review, ix, 3, 34, 68–69, 73, 89, 120, 133, 138, 152
People's Republic of China, research in, 95
Personal Priorities Inventory, 106–107
Pilate, Pontius, 37
Piore, Emanuel R., viii, 77
Poliomyelitis, research on, 42–43
Politicization of science, 96–97, 115–117, 149–150, 156, 158
Press, Frank, viii, 124, 149
Prewitt, Kenneth, 56, 59, 61, 101–103, 105, 150, 153, 158, 162
Price, Derek J. de Solla, 52–53, 55
Priority in science, 136–137
Profession, definition of, 115–116
Protestant Ethic, 112
Public, attentive/inattentive to science, 59, 61, 87–94, 101–103, 105–106, 151, 154
Public Agenda Foundation, 60, 100, 161
Public attitudes to science, ix, 9, 56, 60, 61, 82–97, 100–113, 149–151, 159, 160
Public confidence in social institutions, 85–86, 103
Public images of science, 82–87, 94
Public participation in science policy decisions, 35, 93–94, 101–102, 106, 110–111, 113, 127, 128, 154–155

Radioactive wastes, research on, 9–11
Railroad and the Space Program, The, 58
Rate of adoption of technology, 24
Reagan, Ronald, 122
Reagan Administration, 25, 122–125
Recent Social Trends, 56
Recombinant DNA, research on, 16, 47, 154
Reed, Walter, 39

Index

Research and development (R&D)
  industrial, 22, 38, 70–78
  military, 3, 4, 22, 24–25, 63
  national planning for, 18–29, 126–130
  public sector programs, 25–27
  U.S., 37, 120, 123–124
  U.S., investment, 126–130
  U.S., priorities, 6, 16–17, 152
Resource Conservation and Recovery Act of 1977, 7
Rockefeller Foundation, 39–41
Roosevelt, Franklin Delano, 126–127

Salk vaccine, 42
Sarnoff, David, 37
Sawyer, David Haskell, 60
Science, The Endless Frontier, 3, 38, 123
Science at the Bicentennial, 84, 86
Science education, viii, 79–80
Science indicators
  evaluation of, 56, 58, 161
  use of, 1, 24, 34–35, 54–55
  value of, 48
Science Indicators—1972, ix
Science Indicators—1976, 59
Science Indicators—1980, x, 123
Science Indicators series, x–xi, 159
Science literacy, 15, 83, 96, 128–130
"Scientification" of social institutions, 115
Scientific progress, viii, 123, 134
Scientists' Institute for Public Information, 97
Scientometrics, x
Scientometrics, 52–53, 65
Secrecy
  ethics of, 134–135, 144
  in government, 123
  in science, 131–148, 157
  military, 22, 134, 141, 143–146
  trade, 141–142
Self-governance in science, 127, 130

Shapley, Willis, 61, 76
Shils, Edward, 133
Smith, Bruce, 98
Social indicators, ix, 2, 20, 57
Social Indicators—1976, ix
Social Indicators series, 50
Social inequality, 46
Social invention, definition of a, 58
Social Science Research Council, x
South America, eradication of yellow fever in, 40–41
Stevenson, Earl, 50
Superfund legislation, 7
Surveillance, Epidemiology, and End Results (SEER) program, 13–14

Tax policy, 4, 26–27, 89
Technology assessment, 5, 16
Theiler, Max, 40–41
Thurow, Lester, 54
Toward a Metric of Science, x
Toxic chemicals, research on, 6–9
Toxic Substances Control Act of 1976, 7
Trachtman, Leon E., 97–98
Truman, Harry, 123
"Two cultures," 129

United Kingdom, industrial innovation in, 23
United Nations Educational, Scientific and Cultural Organization (UNESCO), 21, 57
United States Atomic Energy Commission, 9
United States Auto Safety Administration, 11
United States Comptroller General, Report on Science Indicators, 48–49, 53, 59–60, 159
United States Congress, viii, 34–35, 78, 91, 120–121, 124–125, 151, 158
United States Constitution, 96–97
United States Council on Environmental Quality, 19

Index

United States Department of Defense, 63, 90, 121
United States Department of Energy, 121
United States Department of Health, Education, and Welfare
  Food and Drug Administration, 123
  National Institutes of Health, 3, 13–14, 35, 38, 44, 121
United States Department of State, 19, 90
United States Environmental Protection Agency, 123
United States Ethics Advisory Board, 154
United States National Aeronautics and Space Administration, 57, 121
United States National Commission for the Protection of Human Subjects of Biomedical and Behavioral Research, 154
United States National Science Board, ix, x, 34–35, 78–80, 84–85, 90–91
United States National Science Foundation, ix, 3, 38, 68, 78–81, 83–84, 92, 101, 121, 123, 162
  Authorization Act for FY1976, 91
  establishment of, 78
United States National Security Agency, 143–144
United States National Security Council, 90
United States Office of Management and Budget, viii, 35–36
United States Office of Research, 63
United States Office of Science and Technology Policy, 62, 150
United States Office of Technology Assessment, 14, 16, 49
United States Postal Service, research on, 64
Utilitarians, 48

Vernon, Raymond, 24

Walgren, Doug, 158, 161–162
Watson, James, 136
Wealth and Poverty, 123
Weber, Max, 114
Weingart, Peter, 93, 158
Wooldridge Study, 14
Woolgar, Steve, 148
"Working through," 61, 107–108
World War II, secrecy and, 132
Wright, Christopher, 76–77

Yankelovich, Daniel, 60–61, 158–159, 161
Yankelovich, Skelly, and White, Inc., 102
Yellow fever, research on, 39–42
York, Herbert, 63

Ziman, John, 62

www.ingramcontent.com/pod-product-compliance
Lightning Source LLC
Chambersburg PA
CBHW060955230426
43665CB00015B/2205